PRAISE FOR *Erasing Death*

"Parnia is uniquely placed to understand the science underpinning cardiac arrest and CPR and to link this with near-death experience (NDE). More importantly, Sam has the extraordinary ability to communicate these complex concepts in a way that readers will understand. The result is a piece of work that is both stimulating and highly informative."

> —Jerry Nolan, consultant in anesthesia and intensive care medicine at the Royal United Hospital, Bath, UK, and editor-in-chief of *Resuscitation*

"Out-of-hospital cardiac arrest is a leading cause of death in the United States and around the world. Survival after cardiac arrest varies 500 percent from one community to another. Parnia demonstrates that consciousness can persist after the disappearance of any outward sign of brain activity after injury due to lack of oxygen. His story gives hope for future improvements in resuscitation care. Those interested in cardiac arrest should read it and take heed."

> —Graham Nichol, M.D., MPH, FRCP(C), director of the University of Washington–Harborview Center for Prehospital Emergency Care in Seattle, Washington

"Parnia has captured the miracle of life and the ability of CPR to reverse death from sudden cardiac arrest. The medical community and the public should learn from these compelling stories of survivors and celebrate the power of CPR; a skill than can, and should, be performed by every bystander who witnesses another person's sudden collapse."

> —Robert E. O'Connor, M.D., MPH, professor and chair of the Department of Emergency Medicine at the University of Virginia School of Medicine

"In *Erasing Death,* critical-care physician and pioneering researcher Sam Parnia describes in clear language the solid research behind two related revolutions in 'resuscitation science.' First, because we now view dying as a process, we no longer view 'death'—when the heartbeat, breathing, and brainwaves all stop—as final; and recent advances in medical technology now regularly bring people back from death. But second, because we now regularly bring people back from death, we can ask what happens to them while they were dead—and their recollections of that period make possible a scientific exploration of what happens to human consciousness after the body dies, pointing the way toward a new scientific theory of the mind. In finding new ways to save lives, Parnia and his colleagues inadvertently brought the afterlife out of the domain of religious belief and into the realm of science. *Erasing Death* is a solidly researched and carefully presented story that will astound readers and make them rethink what we believe about the border between life and death."

—Bruce Greyson, M.D., Carlson Professor of Psychiatry and Neurobehavioral Sciences at the University of Virginia Health System

"In his visionary new book, *Erasing Death,* Sam Parnia explores life's greatest mystery: what happens after we die? In this fascinating blend of science and philosophy, Parnia convincingly argues that death is a dynamic biological process that only begins once the heart stops beating. As modern medicine continues to blur the transition between life and death, technology has now made it feasible for consciousness to persist in the dying brain, and come back to life. *Erasing Death* takes us on a journey through this new 'window' that lies at the intersection of life, death, and the ultimate nature of human existence."

—Stephan A. Mayer, M.D., professor of neurology and neurological surgery at Columbia University College of Physicians and Surgeons

"What happens when we die, and what lies beyond death's door are among humankind's most enduring questions. *Erasing Death* offers groundbreaking new insights into these vital questions. Parnia is a renowned expert in both resuscitation and near-death experience. Scholarly in content yet remarkably easy to read, you won't want to miss out on this outstanding and highly recommended book."

>—Jeffrey Long, M.D., author of the *New York Times* bestselling *Evidence of the Afterlife: The Science of Near-Death Experiences*

"Based on his studies on 'near-death experiences' (or even better, 'actual-death experiences'), Parnia gives convincing arguments that there is a continuation of consciousness after physical death and that we have to reconsider our current definition of death. An important and highly recommended book."

>—Pim van Lommel, cardiologist, NDE researcher, and author of *Consciousness Beyond Life*

"A fascinating and informative book, written by this cutting-edge researcher on the forefront of resuscitation, provides clinical scientific insights through the study of consciousness during cardiac arrest, which may change the way we practice medicine."

>—Tom P. Aufderheide, M.D., FACEP, FACC, FAHA, professor of emergency medicine at the Medical College of Wisconsin

ERASING DEATH

The Science That Is Rewriting
the Boundaries Between
Life and Death

DR. SAM PARNIA
with JOSH YOUNG

HarperOne
An Imprint of HarperCollins*Publishers*

HarperOne

FIRST EDITION

Library of Congress Cataloging-in-Publication Data

Parnia, Sam.
    Erasing death : the science that is rewriting the boundaries between life and death / Dr. Sam Parnia with Josh Young. — First edition.
        pages cm
    ISBN 978-0-06-208060-8
    1. Cardiac arrest. 2. Sudden death. 3. Cardiac resuscitation. I. Young, Josh. II. Title.
    RC685.C173P37   2013
    616.1'23—dc23   2012041098

13 14 15 16 17 RRD(H) 10 9 8 7 6 5 4 3 2 1

CONTENTS

CHAPTER I

Amazing Things Are Happening Here

JOE TIRALOSI BEGAN TO feel ill shortly after leaving a Manhattan car wash. He was a little nauseated, somehow off, and was glad his shift had ended. A chauffeur, Tiralosi spent his workdays driving legendary stock trader E. E. "Buzzy" Geduld around New York City. But on this August afternoon in 2009, a few minutes after he had begun his drive home to Brooklyn, he couldn't stop perspiring. He cranked up the air conditioner in his car, but he continued to sweat profusely.

Tiralosi was a practical man, a married father of two, and not given to panic. So he planned to push through with the rest of his day, figuring his ill feelings would pass. But an hour later, it was unbearable. He called his wife.

Don't take any chances, she told him. *Go to the hospital.*

But he couldn't drive another block. His wife immediately called a coworker, who found Tiralosi pulled over at the corner of

Eightieth Street and Second Avenue in Manhattan and rushed him
to the emergency room at New York Presbyterian Hospital.

Tiralosi was helped into the ER by his coworker. The color had
drained from his face. He began explaining to a nurse what was
wrong, but before he could finish, he collapsed. A Code Blue, meaning cardiac arrest, was called. Tiralosi's heart stopped. He was dead.

But fortunately for him, he had died in a hospital where a team
of people specially trained in resuscitation was on duty. Doctors
and nurses came racing over from every direction and immediately started CPR. They are accomplished professionals whom I
have worked with many times, including Dr. Rahul Sharma and
Dr. Flavio Gaudio, both very diligent emergency physicians. They
were part of the team that lifted Tiralosi onto a gurney, tore open
his shirt, and cut off his pants with scissors. They attached the cir-
cular electrodes of a defibrillator machine to the skin of his chest.
They moved rolling carts lined with medicines into the cramped
space around him.

Despite all the modern technology available to them, the medical team also scrambled over him with an everyday item—plastic
bags, loaded with ice. They positioned the bags along his sides,
under his armpits, and on either side of his neck. They injected his
veins with chilled saline. The team did all this in about one minute.
His body temperature quickly began to drop. Then they settled into
a rhythm: CPR, accompanied by occasional injections of adrenaline
and defibrillator shocks.

Joe Tiralosi was now surrounded by some of the best medical personnel, technology, and thinking that modern science has
to offer. But he was, with no heartbeat and insufficient oxygen and
nutrients feeding the cells of his brain and body, already dead.

Don't take any chances, his wife said. *Go to the hospital.* Could
these or any other words recur to Tiralosi as he lay flat on the table
and slipped further into the process of death? Was he aware of anything at all? The dominant, scientific view of the brain is that such

a thing would be impossible. The gag reflex and other functions of his brain stem had ceased, meaning his brain had stopped functioning entirely. All the conversations he had with his wife were now seemingly lost to him, and the odds were against him ever seeing his family again.

Seconds passed to the steady rhythm of chest compressions. Minutes passed. They stopped compressions and hit Tiralosi's body with an electric shock. Still, no heartbeat. After ten minutes of continuous chest compressions, the medical and nursing staff was starting to lose hope.

Ten minutes without a heartbeat has long been considered a kind of dividing line in resuscitation science. It has long been thought that after ten minutes without a heartbeat, damage to the brain from a lack of oxygen starts to become permanent. Of course, without a properly functioning brain, Joe Tiralosi would no longer be Joe Tiralosi at all. His memories, his personality, what we might call his "Joeisms" would be gone forever, and only his body would still be here. His wife could hold the hand of the man she had shared her life with, yet they would not really be together.

So ten minutes passed, fifteen minutes passed. Doctors worked well past the old markers; the ticktock rhythm of chest compressions was punctuated by an occasional defibrillator shock.

Twenty minutes.

The call to cease resuscitation attempts in this circumstance belongs to the doctor in charge. But he kept going.

Thirty minutes.

By now, Tiralosi had received thousands of chest compressions and had his heart shocked a half-dozen times. The room was looking more and more like a war zone. Traces of blood and medical debris lay around the gurney. Empty vials of adrenaline littered the floor, like spent gun cartridges on a battlefield. The nurses and doctors providing chest compressions were sweating, consuming their own stored-up energy.

Forty minutes.

Ten years ago, continuing to try and save him at this point would have been considered a tremendous risk—for both Tiralosi and his family. In the best-case scenario, even if Tiralosi's heartbeat was restored, his mind would be a mess—a CT scan likely revealing multiple small and large plumes of damaged, black spaces where functioning neural cells once held his thoughts. But technology and medical understanding have advanced with the years, and so the doctors pressed on because they knew there was a possibility, however remote, that Tiralosi could be saved and returned to his normal life.

Finally, something incredible happened to break the exhausting monotony—someone screamed with excitement: "I feel a pulse, I think we've got him back." Suddenly, in one moment, all those clouds of despair were replaced by a sense of elation in the room. The exhausted staff had a new wind of energy and, more important, after having had more than forty-five hundred chest compressions and having his heart shocked with a defibrillator eight times, and being given countless vials of adrenaline, Joe Tiralosi's heart had started to flicker again.

But the emergency was not over. At this point, precisely why Tiralosi's heart had stopped functioning properly remained a mystery. Doctors needed to find the problem, or there was a very good chance it would stop again. After his heart was restarted, Tiralosi was quickly taken to the cardiac catheterization laboratory, because one of the likely possibilities for his cardiac arrest or death was an undiagnosed heart issue, or more precisely, a heart attack due to a blockage in one or more of the main arteries that supply his heart with oxygen-rich blood. Dye was placed in his arteries to determine if there were blockages.

Frighteningly, while in the cardiac catheterization lab, he lost his pulse again for roughly fifteen minutes—meaning that he actually died a second time. The doctors resuscitated him again. During this process, they found that he had a number of blockages in the vessels to his heart. They opened them with a fairly common balloon procedure and later inserted stents to keep the vessels from closing again. During this entire time, for a twenty-four-hour period in all, Tiralosi's body was kept cooled using a special machine called the Arctic Sun to prevent his brain and organs from suffering damage due to the consequences of a lack of oxygen.

Ten years ago, a man saved after that length of time would most likely have been a kind of living husk—his body present, his mind gone. But today, Joe Tiralosi is a smiling, vibrant man. His face is long and lean with the shade of a well-groomed mustache and goatee covering his lips and chin. He is back at home with his children and the wife whose advice helped to save him, and back at work, continuing his life. The newspapers and television stations that reported on his resuscitation all called his recovery a miracle. If so, Tiralosi and his family were the beneficiaries of a *medical* miracle—delivered through medical science. But to my mind, the word *miracle* seems ill chosen in this context.

Tiralosi was the benefactor of a team of perhaps more than twenty doctors and nurses working in unison using the most advanced medical thinking both during his cardiac arrest and in delivering what has come to be known as "postresuscitation" care. Not only did this bring him back to life, but it stopped any brain damage from occurring. The key component was that the cooling of his body happened in a very timely fashion; it was carried through from the emergency room to the cardiac catheterization laboratory and then continued for twenty-four hours. This slowed the process of cell deterioration in the brain and organs that occurs when the heart is not pumping oxygen. In other words, the processes that naturally take place after death and had started were managed so that he could be revived safely, and most important, he returned to his family without brain damage.

Rather than being a miracle, Tiralosi was one of a growing number of patients resuscitated from death long after we ever thought possible. These cases raise profound questions for doctors, philosophers, neuroscientists, ethicists—and all of us. For starters, although perhaps twenty or so people worked on Tiralosi on this occasion, the reality is that providing this level of sophisticated medical care requires hundreds of people to work together in unison with the mutual cooperation of multiple medical and governmental agencies. Such enormous operations may be commonplace and possible in other industries that require a complex system of coordination, such as aviation, but in medicine, achieving such coordination and teamwork among all the different stakeholders and parties has always proved to be incredibly challenging. Therefore, with so many different people required to work successfully as a team both in and out of hospitals in order to save a patient who has suffered cardiac arrest, how do we ensure that everyone gets optimized care? The painful reality is that even though most of us are not aware of it, many living on our own doorsteps, even in industrialized countries such as the United States, the United Kingdom, or elsewhere, even areas with many of the very best medical centers in the world, may still not receive optimized care. So the big question is, How many more people can we save and how much more can we improve outcomes for resuscitation patients and ensure people do not suffer with permanent brain damage? And then there are the questions where the medical intersects the personal and the philosophical. When does death become final and irreversible? When should people be advised to remove their loved ones from life support for organ donation? What does the recovery of consciousness, after the complete cessation of heartbeat and brain function—or in other words, death—say about the nature of the mind and body or about age-old concepts of the soul and what happens after death—the so-called afterlife? And what further advancements await us?

Those are individual questions, but it is the total picture created by pursuing all these lines of thought that marks the final destination of this book—and the final destination we all share: death.

But the view of death that is emerging may not be one we have encountered. It is one that is at once rigorously scientific, yet also tremendously hopeful.

Throughout history, death has loomed as the ultimate downer of a subject. The ultimate defeat. But recent scientific advances have produced a seismic shift in our understanding of death—challenging our perceptions of death as being absolutely implacable and final—and have thus rendered many of our strongest-held views regarding death as outdated and old-fashioned. In fact, where death is concerned, two major revolutions have already begun—one of accomplishment, and another of understanding. In short, medical science is rendering previously unthinkable outcomes entirely plausible. We may soon be rescuing people from death's clutches hours, or even longer, after they had actually died.

But as an unintended consequence of developing these new lifesaving measures, science is also expanding our knowledge of death. By finding new means to save lives, we are also inadvertently finding new ways to investigate and answer fundamental questions about what happens to human consciousness, to what we might call the mind, the "self," or even "soul," during and after death—questions that, until recently, were considered subjects better suited to theology, philosophy, or maybe even science fiction.

AFTER TIRALOSI'S HEART WAS restarted, he was placed into a medically induced coma for four days, with a ventilator breathing for him. When the doctors brought him out of the coma and removed his ventilator, Tiralosi began telling the nurses that he had a very profound experience. They all recognized that he had recalled something from the forty-seven-minute period during which he was dead.

In popular language, his experience has commonly been referred to as a near-death experience, or NDE. This is a term that I personally don't think entirely and accurately reflects the science of what we are now dealing with, but nevertheless, whether this is psychological or actually happens, these experiences are now reported so routinely that few people who have studied in the field can doubt it is a real phenomenon that warrants further study.

My colleagues called me to hear Tiralosi's story because they know I am involved in a series of studies, all of which revolve around the world opened up to us by resuscitation science. I'm conducting research into optimal cardiac arrest care—the kind of medical science that saved Tiralosi—and into the experiences of consciousness people report bringing back from the other side of death after their hearts have been restarted.

Tiralosi's case raised all the questions I had been studying. When he was lying on the table with no heartbeat, where was his true self, his mind and consciousness, his memories? Was he aware of what was happening to him? The dominant scientific view is that he had entered an abyss of experience—the sunless void of existential nothing.

I met with Tiralosi in his hospital room a few days after he awakened from the coma. A tall, slim, middle-aged, gray-haired Italian American man, Tiralosi took a few moments to gather his thoughts. His wife held his hand and looking at him lovingly as he gazed at a small yacht that was floating across the gently rippling waters of New York's East River and told me his story.

What gripped me is that he recalled only one detail during the time his heart was not beating, but what he remembered affected him profoundly. He said that he had encountered some sort of spiritual being, though nothing that had mass or a shape. He described encountering a luminous, loving, compassionate being that gave him a loving feeling and warmth. His encounter with this being was ineffable. He couldn't find the right words to fully describe his sensations. This encounter and the whole experience had comforted him to know what it would be like when, in his own words, he goes "to the other side." Because he had experienced this luminous feeling, he said that he was no longer afraid of death. Whatever this being or feeling was, it completely transformed him.

On the surface, this seems to be a truly remarkable reaction for any person who had such a close brush with death. As a critical care doctor, I see what becomes of people who don't report this conscious recollection of death. While they often express some sense of relief at having survived, they may become physically, mentally, and emotionally fragile. Life has impressed its tenuousness upon them, and it has warned them that death is real, not something that happens to other people. This raw sense of their own mortality can be difficult to endure, and long-term psychological disorders such as posttraumatic stress disorder and depression are not uncommon in resuscitation patients.

However, people who undergo an experience like Tiralosi's truly seem to be in a new world—one in which death, for them, is nothing to be afraid of. Tiralosi felt he had met with a luminous being, and he came away from the encounter with a new understanding of his role as a husband, friend, and father. Like others who have reported experiencing such an event, he came out of this feeling less materialistic and more altruistic.

His story is, in a sense, personal to all of us, as it speaks to some of these fundamental mysteries of human existence. But it is particularly personal to me—because his tale fits into my area of research, and because the hospital where Tiralosi was saved by medical science is the hospital where I worked until very recently. Hospital administrators there advertise with the slogan "Amazing things are happening here." Of course, no one considers his or her own workplace to be necessarily amazing. But the truth is, amazing things *are* happening in medicine—things that suggest our own lives and consciousness might be even more amazing than science has thus far allowed.

Different groups of doctors and researchers in various fields are forging this new path. They are creating cold packs to enwrap heart-stopped patients, machines to push chilled saline into their veins, injections to preserve the cells of the body, and drips that deliver oxygen carefully draped by a microscopic layer of fat to cells in the remotest parts of the body after death, as well as equipment to deliver more effective chest compressions. Neurologists have started to discover that contrary to the old dogma, supposedly vegetative patients may actually be conscious and aware of their surroundings and can carry out mental tasks on command.

I have authored one of the first medicine-based studies ever con-ducted of what happens to the mind and psyche after the heart has stopped beating. Currently, I am conducting the world's largest ever study of mind and brain during cardiac arrest, the AWARE study, an initial three- to five-year-long investigation of patients who claim to have been aware of being resuscitated. AWARE received world-wide press and was announced at a conference sponsored by the Nour Foundation,* the United Nations Departments of Economic and Social Affairs, and the University of Montreal in September 2008. The conference itself was something of a paradigm-busting event—as it showed that consciousness studies and resuscitation sci-ence are intertwined, and their implications bear on us all.

THE CASE OF JOE Tiralosi illustrates the beginning of what resuscita-tion science can do. It also shows that determination of how long to

---

*As a public charitable and nongovernmental organization in special consultative status to the United Nations Economic and Social Council, the Nour Foundation's mis-sion statement is to explore meaning and commonality in human experience through a multidisciplinary and integrative approach that seeks to engender greater understanding, tolerance, and unity among human beings worldwide. The conception of the founda-tion was inspired by the inclusive philosophy of the late Ostad Elahi (1895–1974), a thinker, jurist, and musician.

The foundation provided us with a grant for the AWARE study, which was launched during its symposium at the United Nations in New York on September 11, 2008. For more information, visit www.nourfoundation.com.

resuscitate is purely a subjective one and that conforming standards are needed to make it objective. Undoubtedly, in a different hospital or even in the same hospital but on a different day and with a different team of doctors, Joe would not have been revived for that long. The team would have stopped long before the forty-seven minutes needed to bring him back to life, and he would not have received the cooling therapy and other vital care that he received, such as a timely cardiac catheterization that took place immediately after his heart had restarted in the emergency room and continued even while his heart had stopped again a second time. If any of these and other procedures hadn't happened, Joe Tiralosi would either have been permanently dead or would have possibly been permanently disabled or in a vegetative state.

The cooling technique that saved Tiralosi has opened the door to a whole new field of research, proving we can buy time to correct all manner of medical disturbances and still recover a *whole* person, with both physical and cognitive functions intact. However, it has been estimated that just 50 to 60 percent of hospitals in industrialized countries, including the United States, Britain, and Germany, employ the cooling procedure that preserved his body and mind.

Other cutting-edge research is under way. Dr. Robert Neumar and others, whose research I'll address in greater detail later, are working on a pharmacological solution that may help preserve the body at a cellular level, prolonging the body's natural hibernation phase, while doctors fight to revive the patient. And the miraculous little extra-corporeal membrane oxygenation (ECMO) machine allows health-care providers, particularly in Southeast Asia, to reroute a patient's blood out of the body, feed it oxygen, and return it to the circulatory system. These are truly game-changing medical advances that threaten to topple all our long-held ideas about death and its power, that are poised, in fact, to give humanity more power over life and death than ever before.

Tiralosi also had a profound experience. This dovetails with the philosophical issues. What does the science of resuscitation say about consciousness, and the ability of the human mind, consciousness, and soul—or, in other words, that entity that makes me into who I am—to survive death? And what, in turn, does that say about the relationship between mind and brain? The answers to these questions are of course profound, with implications for science, philosophy, religion, and every man, woman, and child. We are just beginning to explore these answers in a society where medicine and religion try to coexist but are often at loggerheads. The mystery of what happens when we die is something everyone pauses to consider, a question to which we would all like a definitive answer.

As we explore the philosophical and scientific byways surrounding the mysteries of human consciousness, we will try to consider what all this means for how we approach the subject of death, how we pursue further scientific research, and perhaps most important, how we interact with one another.

I BECAME INTERESTED IN resuscitation science when I was twenty-two years old and have devoted my working life to it. Currently, I divide my time between hospitals in the United States and the United Kingdom. I am an assistant professor of pulmonary and critical care medicine and the director of resuscitation research in the Department of Medicine at the State University of New York in Stony Brook. I obtained my medical degree from the University of London before completing my specialist training in internal medicine, pulmonary medicine, and critical care medicine at the Universities of Southampton and London in the United Kingdom and Weill Cornell Medical Center in the United States. I was also awarded a Ph.D. in cell biology from the University of Southampton in the United Kingdom.

A combination of different events and questions drew me to this subject. The first thing that sparked my interest was studying the brain in medical school. One day in the neuroscience laboratory as we learned the function of the brain, I was awestruck and wondered how this incredible gray organ could lead to all our personalities and everything that makes us unique as individuals. One of my friends at medical school was very introverted and hardly ever spoke. I remember looking at her one day and thinking, *What is it that makes her so different from the rest of us?* Then I looked around the room. There were fifty people, and though we shared many similarities, we all had unique personalities. What was it about this organ that made us all so different? Where did our mind and consciousness or, as the ancient Greeks called it, the "psyche" or "soul" come from?

Toward the end of my medical training, I came across people who died. I began contemplating what happened to the minds of these people who died. I also noticed that there was little and limited science involved in the decision of whether or not to resuscitate a patient. It was not objective enough; it was actually purely subjective. In those days, doctors often didn't even ask patients when they were admitted if they wished to be resuscitated or explained what it meant. The doctors would just make a decision and write on the chart "do not resuscitate."

All this was developing in my mind and finally culminated when I met a patient in the emergency room. I was twenty-two and in New York as part of my final year of medical studies on a clinical attachment at Mount Sinai Hospital. It was an exciting time for me. I was at one of the finest medical institutions in the world, working my way through medical adolescence and into adulthood.

One morning I was making my rounds when my pager vibrated. I rushed to the emergency room and picked up the notes from the nurse, which read: "Desmond Smith, hemoptysis"—a medical term for coughing up blood.

Desmond was a tall, thin man of West Indian origin with a distinct Harlem accent and a winningly contagious personality. Most patients I see in the emergency room are understandably complaining about pain and upset at their condition, and therefore aren't given to small talk. However, sometimes out of nervousness, or sometimes just out of inherent friendliness, patients share details of the day-to-day events in their lives. I quickly found out that Desmond was one of the friendlies. He told me that he was sixty-two years old, that his family had recently thrown him a surprise birthday party, and that he was not at all concerned about his health.

As I palpated his chest, carefully searching for any sign of abnormality, I learned that Desmond began his day with what had become a daily bout of early morning coughing. Carrying his breakfast tray to the bedroom, he recalled his doctor's original comment: "It's a smoker's cough." But that day, for the first time, Desmond coughed up blood.

Still, Desmond was optimistic. "That's what I coughed up. Never mind. I'll live, Doc!" he announced.

I detected signs of fluid surrounding his lungs, so I ran through a mental list of possible diseases. The most common cause of coughing up blood is a simple upper respiratory tract infection, a flulike illness. But this didn't seem to fit Desmond's case. He was a lifelong smoker, so there was a chance he had lung cancer. His vital signs were strong, so I decided to order further tests. But whatever Desmond had, it didn't seem to be a life-threatening emergency at that moment. Desmond gave a brief thank-you salute, and I left the emergency room.

Less than thirty minutes later my pager went off again, declaring: "Cardiac Arrest: Emergency Area. Cardiac Arrest: Emergency Area." This was life and death. As I ran down to the emergency area, adrenaline rushed through me. When I arrived, a bay had been curtained off. I pulled the curtains aside. A team of doctors was frantically working on a man. One was kneeling by his head, hurriedly trying to secure his airway. There was blood everywhere. Time sped up for me as I realized I knew this man: it was Desmond.

There was a frantic rush to save his life, with doctors shouting orders in rapid succession. *"Pulse check, rhythm check . . ."* *"VF . . ."*

"Shock . . ." "Stand clear. Oxygen away!" Thud, thud. "Get intravenous access. . . ." "1 mg epinephrine, stat." "Continue with CPR. . . ." "Start a bag of fluids." "Blood's pouring out of his mouth, he's bleeding extensively. . . ." "Suction, quick!" "Get the double lumen endobronchial tube. Get the emergency bronchoscope. We've got to find the bleeding vessel. . . ." "1 mg epinephrine stat." "Cross match." "Universal blood stat." "Squeeze the bag of fluids. . . ." "Asystole . . . flatline . . . 1 mg epinephrine, 3 mg atropine stat." "Continue resuscitation." "I can't see anything—it's just a red sea of blood down there. . . ." "It's impossible to resuscitate him; he's clotted up his airways. . . ."

Just like that, Desmond was dead. One minute he was here, the next he was gone. What had happened to the person I had been talking to a half hour ago about his surprise birthday party? What was left of his memories, thoughts, and feelings? It appeared there was just a lifeless body.

This interval between life and death had been so quick. Questions buzzed around in my head. What had Desmond experienced? Had he been able to see us trying to resuscitate him? What was happening to him now? Could he have retained some form of consciousness, or was that the end? Even with my medical training, I couldn't even begin to answer those questions.

The death of Desmond and the questions I kept asking myself about the process he went through in those minutes and possibly beyond had a profound impact on my life. It was so deep that in the coming months I decided to pursue the answers to these mysteries through the tool I had begun to learn and could rely on the most: science.

This subject, which started out as a point of interest at medical school, grew with me at different stages of my own medical development. As I witnessed decisions being made about life and death for patients, even as a young medical student I realized that we needed an objective science. I then graduated from medical school and decided to find out for myself what happens to patients. While setting up a study at the University of Southampton during my residency training, I also set up a separate study where I collected approximately five hundred cases of people who had what were called near-death experiences under different circumstances. This taught me a lot about the nature of the experience and its impact on people.

I began to see that the people who had these experiences were from all different backgrounds and all different belief systems, ranging from agnostics to atheists, and from people with a minimal religious predilection to strongly religious. What was most touching about the experience was the fact that for many people, especially those who had encountered a being of light, as they described it, they had been profoundly affected and were transformed positively by the experience. The other thing that struck me was how physicians and nurses had often been involved with a resuscitation of a person who had such experiences, and to their astonishment the patient had come back and told them in detail what had been happening even though the person appeared to be dead.

As I further developed my medical practice, I began to notice that learning how to save a life involves learning all the different components and links in a very long chain of survival. This therefore becomes the science of resuscitation, and what became more and more obvious to me was that if attention was not paid to all these links in the chain, then patients that we take care of in hospitals might experience more of an adverse effect, including higher fatality and long-term brain and other organ damage. Furthermore, as I took a keen interest and specialized in resuscitation science it became very clear to me that although individual doctors and nurses did strive to provide the best care possible, even more could be done. But the reasons they weren't always being done was largely a system-based issue that needed to be addressed at a level far above and beyond individual nurses or doctors like me. Through questioning why patients have near-death experiences, I came to eventually realize how little clinicians actually know about the quality of resuscitation with

respect to the brain and other vital organs during the period of cardiac arrest. It suddenly dawned on me that we had been "driving in
the dark" for years without a real-time gauge to tell us whether our
treatments and interventions were being effective, like a driver who
would only know whether he had been successful in driving if he
arrived at his destination but with little information in between. In
the same way, the only way we would know whether we had been
successful in resuscitation would be whether someone like Tiralosi
survived or not. If the person didn't make it, then we would all
put it down to the inevitable "he had crashed" and we hadn't been
able to reverse death, because it was death. But with time it became
clearer to me that actually in many cases permanent and irreversible
death was not inevitable even if death had taken place. It was simply
that in spite of the best efforts of resuscitation doctors, somewhere
along the complex chain of survival needed to bring someone back
to life, one or two links had not been in place. This raised the
question as to whether these experiences and recollections from the
period of death that some people recalled could simply be telling us
that certain people had better quality of resuscitation of the brain. If
that was the case, then clearly they warranted further investigation
so we could understand what we were doing that was leading to this
improvement. These experiences could also be telling us something
more about the philosophical and personal questions that we have
all had about what happens when we die. This, of course, is only
possible because we now know that death is reversible.

CHAPTER 2

One Small Step for Man, One Giant Leap for Mankind

T HERE WAS A TIME when space exploration was viewed as impossible. Roughly a hundred years ago, people would have thought you were mad if you proposed a mission to the moon. They would have asked how in the world you could send a person into the vast unknown and return him or her safely to Earth. When the topic was explored in books, it was placed in the realm of science fiction. In 1901, H. G. Wells, the renowned author of *War of the Worlds,* published *The First Men on the Moon,* a story about two men who build a spaceship and travel to the moon. Though his novel was categorized as pure science fiction and regarded by some as preposterous, Wells was convinced that space travel would one day be possible.

Not only is space exploration now possible, it is taken for granted. Because of the advancements in science, we were able to witness the first successful moon landing in 1969, which was a new beginning for us all. By analogy, it is the advancements in science that have allowed us to cross over into the boundary of death and explore it. This is the crux of resuscitation science—the science of bringing people back to life after death. This may sound impossible, as if we were veering into science fiction territory, but it is not. It is very real.

To this day when we talk about death, people have the same reaction as our forefathers did one hundred years ago, around the time of the publication of Wells's novel.

Because of the pace of medical progress, few doctors, and even fewer nonmedical or scientifically oriented people, understand that a true revolution has taken place at the height of human understanding that will have huge implications for all of us irrespective of culture, creed, or background. This is truly a global phenomenon. Contrary to popular social and even medical perceptions, "death" is not the end we once thought it to be. Death is no longer a specific moment in time, such as when the heart stops beating, respiration ceases, and the brain no longer functions. That is, contrary to common understanding, death is not a moment. It's a process—a process that can be interrupted well after it has begun.

Although it is true that death is a biological process and that people may die as the result of multiple causes such as an infection, heart attack, or cancer, the end result of the whole process from a medical, biological, and cellular perspective is that there is some process that leads to inadequate delivery of oxygen and nutrients to the cells in each organ. These are required to sustain the activity of the organs and cellular processes that keep us all alive. When this process happens to the heart, it stops working. There is no longer a system to pump blood around the body, the organs become deprived of oxygen and stop working within seconds, and we are then lifeless. This was the end—or at least this is what has been considered for millennia to be the end. Incredibly, it now doesn't seem to be. Since the promotion of resuscitation science in the 1960s, doctors have regularly restored the heartbeat after death and, according to the common lexicon, *brought people back to life.*

The reason death can be reversed is that it's a process, not a moment. Death is, biologically speaking, a stroke, but unlike an everyday stroke, it is the entire brain that has become deprived of oxygen and nutrients; and the process that transitions brain cells from being in a potentially reversible to an irreversible state of damage and cellular death goes on for many minutes to hours after death has started in a person. In a common stroke, there is a lack of blood supply to a part of the brain, usually but not always as a result of a clot. In death, due to the heart not beating, there is a lack of blood supply to the entire brain rather than just a local area. Although the cause of the lack of blood supply is different between a stroke and death, as far as affected brain cells are concerned, the same biological processes occur when they are starved of oxygen. In a stroke, the longer the blood supply is interrupted, the greater the chance of permanent damage to that part of the brain. We know that even if someone has a stroke and you get the person to a hospital equipped with the right technology in the first few hours and do a CT scan, then a doctor can try to go in and immediately open up that blood vessel. If the doctor can open that blood vessel, most of the brain tissue can be salvaged, and the person can recover and be restored to a high or even full quality of life. But if this is left for too long, the person will suffer permanent and irreversible brain cell damage and hence a disability. If we intervene even after death has started, we can make a difference, and this is why the goalposts have moved in terms of understanding death and the processes that occur after death.

So if you think of death as a stroke, then you realize that, like a stroke, it can be reversed. This has significant implications for how death is defined. If someone says, "My aunt died," we may innocuously ask, when did that moment happen? The truth is, from a scientific perspective we may not know anymore. And if someone asked me when exactly death becomes permanent, again I would have to say we don't really know. But we do know that the once-held philosophical idea that there is no way back is not accurate and that for a significant period of time after death, death is in fact fully reversible.

Since there is a way back from death, we must now ask what happens to a person's mind and consciousness—that very essence of who the individual is—during the period when the person enters that unknown territory. Since there is a fairly long period during which what doctors do (or don't do) will have implications for the patient's quality of life, it's important to understand this new "gray zone." The treatment that we administer or neglect to administer during this period of time can mean the difference between someone returning to a meaningful, purposeful life or someone being in a vegetative state. And finally, there is the philosophical concept of what happens after death or during the period of a so-called afterlife. Since, by definition, the person has passed over the threshold of death for what can be as long as a number of hours, he or she is in what some have in the past called the period after death. If the patient is brought back to life, what can the person tell us about death and what happens when we die? This is where objective medical and scientific advancements aimed at saving lives and brains have inadvertently intersected with people's personal beliefs, their religious or overall worldview.

Admittedly, these can be challenging concepts to grasp, largely because our concept of death has traditionally been very black and white. In movies, we see very clearly when people are dead. Someone shoots someone else several times, the guy falls down, and he's dead. That's why when someone became lifeless and motionless, and the individual's heart had stopped, classically the person was considered dead. In fact, physicians still use these simple three criteria to declare someone dead: no heartbeat, no respiration, and fixed, dilated pupils (which means that there is no brain function).

Frequently, people—including doctors and scientists—say to me that if someone was brought back to life, then he or she wasn't actually dead. How could the person have been dead if the individual is now alive? But the fact of the matter is that death isn't what you and I decide it to be. Someone may have a philosophical idea that death is defined as when a person cannot come back and talk, eat, or share stories. That may have even been our own belief or our doctor's belief, but that's not death. People don't set the parameters for what death is; science does.

Let's go into some detail on what death actually is. Biologically and medically speaking, "death" and cardiac arrest are synonymous. By definition, cardiac arrest is when the heart stops beating and the person stops breathing. Within seconds the brain also shuts down due to a lack of oxygen and nutrients being delivered to it, and the patient develops fixed dilated pupils. The medical term for death is cardiac arrest because at this time all three criteria of death (i.e., no heartbeat, no respirations, and fixed dilated pupils) have been reached.

Many equate a heart attack with cardiac arrest, but the fact is they are not the same thing. Someone might say, "My uncle had a heart attack and lived; therefore, he was brought back from the dead." But this isn't correct. A heart attack occurs when an artery supplying blood to the heart becomes blocked, thus preventing the heart from receiving proper blood supply. As a result, part of the heart muscle dies. A cardiac arrest is when the heart stops beating for any reason, which signals death. Of course, a large enough heart attack can also cause the heart to stop by blocking blood flow into the heart itself and thus lead to cardiac arrest, as occurred with Joe Tiralosi. But unlike a heart attack, cardiac arrest and death are medically defined as the same thing; there's no difference.

Doctors often believe that in the case of cardiac arrest, the course of action is to administer emergency resuscitation measures, the most basic form of which is cardiopulmonary resuscitation (CPR), to try to bring the person back within minutes. If they cannot, then the person is pronounced dead. When modern resuscitation was first discovered in 1960, it primarily involved just chest compressions, breathing, and the ability to electrically shock the heart, and this is still what most people associate with resuscitation science. But today science has progressed far beyond this.

Now, after longer and longer battles with death that extend hours after death has occurred, doctors are able to save more people after cardiac arrest than ever before. What has been discovered is that when a person's heart stops beating, when the crucial supply of oxygen and nutrient-rich blood to the heart and all other organs including the brain has seized and the person is declared dead as we know it, all the cells in the body do not immediately die. Brain cells, liver cells, and muscle cells all have a period of time after the heart stops and a person dies before they become irreversibly damaged. They have just begun their own process of death, which can take hours. For example, scientists can take sections of the brain from someone who has died four hours earlier and been taken to a mortuary and grow the cells in the laboratory, indicating that those cells were not irreversibly dead. Though the cells are not working, they are still potentially viable. This is why people can donate their organs and those organs can then be used in a healthy body. To function, cells require oxygen to be delivered so metabolic activity can take place and energy can be produced. Without oxygen, those cells start to go through a process where they will die, but they haven't died yet and they can still be saved. However, during this time, that organ is not working because its cells are not getting any blood or oxygen. So although the brain is not functioning and the liver is not functioning after the heart has stopped, the process of cell death in those organs has only just begun and can go on for hours after a person has been declared dead.

So where along this process do we draw the line and say a person has reached the point of permanent death? Anywhere we draw that line and say death is permanent and final is going to be somewhat arbitrary, because death is a process and the line at which certain cells and organs permanently die will be constantly moving as the science of resuscitation progresses further and further. The further we can extend it, the further we can extend that final, absolutely fatal point. Much of the new frontier in medical research is focused on prolonging the state where the cells are still viable and buying time to reverse whatever underlying condition caused the cardiac arrest.

There are two consequences to resuscitating people—just as there are two processes involved in space travel. First, if the cells are salvageable, not only can we bring people back to life from the clutches of death after they have gone down into what was thought to be that eternal abyss, we can also ensure that they don't end up with any permanent damage—like the astronaut who has gone to space in a rocket and returned in a capsule. Second, in the same way that an astronaut is exploring space and telling us firsthand what is going on there, these people can tell us what is going on in the metaphorical space of the afterlife. People all over the world have developed their own ways of thinking about this, in much the same way that when thousands of years ago people looked up to the sky, they wondered what was really going on in space. But these beliefs are often personal beliefs. The same is true of death.

Bringing individuals back to life is only the first step in survival. If someone dies and we bring the person back, doctors also have to fix the underlying problem if the individual is going to remain alive. If someone suffers a cardiac arrest and we can rapidly treat the underlying condition, then death can be reversed. For example, if someone is in a motor vehicle accident and is bleeding heavily due to a ruptured blood vessel and then has a cardiac arrest and dies, there are ways to save the person. This is a blood loss issue. Whereas many may simply accept defeat and declare the person "dead," now we could potentially set about establishing a system that slows down the process and rate by which cells in this dead person's organs are becoming permanently damaged and thus prevent long-term brain or other vital organ damage. By doing so we buy our expert surgeons a few hours of time to find and treat the bleeding vessel, replace the blood that was lost, and then restart the heart—and therefore reverse death. If a person is in the hospital and has a heart attack because of a clot that has entered an artery in one side of the heart, doctors can rush the person to the cardiac catheterization lab, open the patient up while the individual is dead, remove the clot, and then bring the person back safely. A sophisticated bundle of timely medical interventions, including what has become termed "postresuscitation" care delivered correctly in the subsequent seventy-two hours or so after the heart has been restarted, can ensure such a patient can even resume a normal life.

To me, this is even more amazing than going to space and returning. After all, how many people directly benefit from space exploration? Yet everyone benefits from the advances in resuscitation science. And, in the same way astronauts can explain what happens in space once they return, these people who have died can tell us what they experienced during the first few hours of death.

Clearly, bringing people back has limits, in the same sense that how far we can travel in space and return safely has limits. In the event of a horrific accident that destroys all of a person's organs, we can do nothing for that person because there are no organs to support him or her. There has to be something to come back to. If the body is completely destroyed, it can't be rebuilt—at least not today. However, those limits are continually being pushed back. In the same way that we first went to the moon and now we are trying to go to Mars, the same is also true with resuscitation science. We are pushing back the boundaries of life and death and extending into the period that had been considered the afterlife. Today, it is three or four hours; tomorrow it could be twelve or twenty-four hours and beyond.

An active tissue regeneration program, where scientists are working diligently to develop systems to grow artificial organs, already exists. In July 2011, doctors in Sweden created an artificial trachea from a man's stem cells and replaced his cancerous windpipe with the synthetic one. One can imagine that, say, in one hundred years, if this technology has been fully developed, then we may be able to rapidly implant new organs and reconstitute a person who had irreversible damage. Today, if someone has cancer for which therapies are available—but unfortunately before we got rid of the cancer, that person crossed over the threshold of death—the patient can be brought back safely, and if the person has a potentially curable cancer, he or she can then receive therapy and get better. Of course, that's not the case if the person has, for instance, terminal lung cancer, because at present, there is no cure for terminal lung cancer or certain types of resistant lymphoma or leukemia that have become refractory to treatments.

In 2011, I treated a courageous twenty-seven-year-old woman named Laura who had leukemia. People suffering from leukemia experience a proliferation of their white cells, which are called blast cells because they are not fully formed. These cells frenetically replicate, and the proliferation of the cells that don't work causes damage to the organs. If the cells were mature, they could fight infection, but at this stage, they end up just clogging the system.

Laura had been diagnosed with leukemia at age seventeen. Though leukemia is sometimes a chronic condition, it can in some instances be cured with chemotherapy or through a stem cell transplant. Laura had in fact received a stem cell transplant. In this process, she was given such strong chemotherapy that her bone marrow was almost completely annihilated, which essentially stopped her body from producing the bad white blood cells. Doctors then transfused stem cells from a family member, and these new stem cells produced healthy white blood cells. Though the procedure can have complications, such as an infection, it worked on Laura, and she was cured.

For the next ten years, Laura lived a normal life. She married and had a daughter. Though Laura never forgot the leukemia and the grueling treatments, it was something that haunted her past rather than guided her future.

Then, like a delayed aftershock in an earthquake, at age twenty-seven, her symptoms returned. She went back to her oncologist. Just as he had done the first time, he gave her multiple doses of chemotherapy and arranged for her to undergo another stem cell transplant. Unfortunately, it didn't work the second time, and her leukemia became refractory, meaning it did not respond to treatment. The oncologist told her and her family that there was nothing else that could be done. But because she had a young daughter who was the jewel of her life, Laura wanted to do everything possible to prolong her life.

This was when I became involved as a critical care doctor. When I first saw her, Laura had physically deteriorated and suffered significant weight loss. She had contracted pneumonia because she didn't have a functioning immune system to fight the infection. Worse, the pneumonia was completely resistant to all antibiotics and continued to grow because she had no functioning white cells to fight the infection. She just had "blasts," which were damaging all her internal organs. She had reached the point where she could barely breathe. Based on her wishes "to do everything," I had a breathing tube inserted and tried to continue addressing the infection, but this was difficult because it was not responding to any treatment. Because the infection wasn't clearing out, Laura became progressively worse.

It became clear that her heart would soon stop. In theory, based on everything we have covered, a logical question would be this: Knowing her heart was going to stop and having time to prepare for that moment, why couldn't we just reverse her death when that happened? The fact is that I knew we could have restarted her heart, but the real problem was that it would stop again and again because her organs were not getting proper oxygen delivery. We didn't have the science to stop the underlying problem that would cause her heart to stop. We couldn't get rid of the blast cells and ensure she had normal white cells. Her oxygen requirements had skyrocketed due to the infection. We had tried to match those requirements by putting her on the breathing tube, but because she was getting sicker and sicker—due to an untreatable infection and ongoing organ damage from the blasts—there would come a point at which we could not match her oxygen requirements. We just couldn't keep up. This would cause her blood pressure to drop, and as that happened, she would not be able to deliver the oxygen to her organs that we were giving her through the ventilator. She would soon go into shock, causing her organs to enter a vicious cycle ending in her heart stopping: cardiac arrest and death.

Because of all the inflammation in her body, Laura had a huge bag of fluid inside of her body that had collected around her heart, and it eventually caused her heart to stop. In a normal cardiac arrest, we could insert a tube in and drain that fluid, but in her case, because of the underlying issue, this would not have saved her life. Her family made the emotional decision to stop treatment, and sadly she died. We didn't even attempt to reverse her death. We had lost that battle a long time before her heart had stopped. So while we can reverse death, it is only meaningful if we have the means to reverse the underlying problem. Laura's case illustrates this. Even though my colleagues and I could have restarted her heart, we knew that it would stop again because the disease process was pushing her rapidly downward. This is what people need to understand. Doing "everything" sometimes actually involves understanding where medical care may simply be futile.

Therefore, the limiting factor is not the process of death itself— this can be reversed and managed—it is the process that leads to death. For example, without the right antibiotics to treat the *E. coli* infection that engulfed Germany in 2011, people would have died in great numbers. If there is an overwhelming infection that causes the heart to stop but for which a treatment has been discovered, then we can bring people back to life safely and restart the heart through resuscitation science even after death and put them on the path to a full recovery. The same is also true of cancer. For some cancers we have discovered appropriate therapies, but for others, we haven't. So while we can restart the heart of a person with cancer, if it is at an advanced stage and it is this that is causing the person's heart to stop (as with Laura), even if we manage to restart it, it will stop again and again because the disease is out of control and we have no way of treating the underlying condition, the one that is causing the heart to stop. Clearly, as science and medicine progress, we will find better tools to fight causes of death, so in a sense, there are events that lead up to death and there is reversing death itself.

Almost every doctor working with the critically ill in emergency rooms and hospital wards has a resuscitation story to tell. I myself have many such stories. Recently, in fact, as I was walking through the hallway at Stony Brook Medical Center on morning rounds, a woman came racing into view, screaming.

"I think she died!" she yelled. "I think she just died!"

The woman was speaking of her sister, a patient in the intensive care unit. I quickly bolted into the room from which this upset woman had just emerged. The patient in question, whom we'll call Carrie, was in her twenties, suffering from a chronic kidney condition. She had been admitted to the hospital with a severe infection.

I must admit that, despite my haste, I entered the doorway feeling a bit skeptical that the situation would turn out to be so urgent. This young woman had been on the mend, and she was due to be discharged from the intensive care unit soon. However, my first look at Carrie confirmed her sister's concern; Carrie's pallor was gray and turning blue. Her heart had stopped. She was, by all the standard definitions, dead, just as Tiralosi had been.

By this time, mere seconds after being called, the full complement of staff we needed was in the room, ushering family members out and getting to work. We began administering chest compressions, followed by a shot of adrenaline. That did it. Carrie was back—just as fast as she had gone. Her color returned, and soon her family was back at her bedside. Though I was the doctor in charge of her resuscitation, I felt much like a witness—a bit awestruck at the scene. That old-fashioned moment of death had been efficiently brushed aside.

Carrie's case offered nothing profound in terms of what she experienced during cardiac arrest. Many cases like hers do not. However, in Joe Tiralosi's case, I saw the convergence of the two lines of investigation I'm pursuing—the science of medical resuscitation, and the larger, more fundamental mystery of human consciousness.

The headline here is that death itself is not the problem. That doctors can reverse. If a medical team has all the equipment they need and the proper training, they can make sure that many of their patients who suffer cardiac arrest and die are brought back safely and don't get brain damaged. But obviously we need everyone to better understand the implications of resuscitation science. If we devote enough resources to resuscitation science, then we can make sure many more people who have a cardiac arrest and die can be brought back to life safely and have meaningful lives instead of suffering the consequences of damage to the brain and other vital organs. Someday, one of those lives will be yours and mine, or our mother's, father's, or children's. It has been estimated that over one million cardiac arrest cases occur in Europe and North America per year. This means deaths where resuscitation was attempted. One can only imagine how many cases occur around the world. Therefore, a small fractional improvement will lead to hundreds of thousands of lives and brains saved and an enormous reduction in the health-care burden and costs to take care of people with long-term brain injuries. Perhaps we were never meant to conquer gravity and fly or even go into space. But in the same way we have done so, we can also go past the point of death—and therefore we need to work to ensure a safe return.

Here are some numbers to consider. In the United States alone, experts estimate that more than $5 billion a year is spent on cancer research and treatments—and an estimated $5 to $7 billion of NASA's budget was spent on space exploration in 2011—but very little is spent on research in the area of resuscitation from cardiac arrest. But there should be, because, as my colleague Dr. Charles Deakin at the University of Southampton in England has so eloquently stated, cardiac arrest is the one thing that is going to happen to every single one of us, unlike cancer or other conditions that only affect some people. Not properly researching cardiac arrest undercuts all other forms of medical research. If a cancer patient turns out to be curable and is treated but suffers a cardiac arrest, and we don't bring that patient back safely, then all the effort that had been put into the cancer treatment has been wasted and a life is needlessly lost.

What's really unfortunate is that if we look at survival rates from cardiac arrest, there are tremendous variations from place to place and hospital to hospital. Furthermore, even in the same exact hospital, survival figures are likely to be far worse on weekends and evenings. It has been reported that where I work, if someone suffers a cardiac arrest outside of a hospital setting and receives CPR, and is taken by ambulance to the hospital, the person's chance of being successfully revived and surviving to discharge is only approximately 2 to 3 percent per year. This is one of the lowest figures not only in the United States but also probably the world since there isn't much room to go below 2 or 3 percent. The statistics for in-hospital cardiac arrests are a little better, but still not great. In the United States and United Kingdom, the overall average survival to discharge for this group has been estimated at only 16 to 18 percent per year. However, many of the academic medical centers such as ours can achieve 21 to 24 percent survival levels, which implies that many other hospitals must be achieving survival levels well below 16 to 18 percent. As we shall see, the reasons for the discrepancy in survival figures across different communities and hospitals even in the same country are complex. Furthermore, translating the latest research discoveries into patient care has always been a major challenge to all health-care providers. Add to that the fact that, by virtue of having died and being revived back to life, this somewhat "unnatural" phenomenon means that by definition those patients are the most critically ill patients in the hospital and require enormous resources as well as very specialized skills that may not always be recognized and thus available. The reason for the discrepancy between in-hospital and out-of-hospital cardiac arrest survival levels mainly reflects the fact that while patients in the hospital are likely to be more seriously ill than those in the community, expert care can be administered more quickly in the hospital than out of the hospital. In Richmond, Virginia, where they also started out with an approximately 2 to 3 percent out-of-hospital survival rate, measures were put into place step-by-step and over a number of years by diligent doctors such as Dr. Mary Ann Peberdy and Dr. Joseph Ornato such that in that community, survival rates increased to as high as 18 percent. That means for every hundred people who suffer cardiac arrest, they now routinely bring back fifteen or sixteen more people per year. Although this is clearly very impressive, as we will see later, some communities have even more impressive survival figures. With respect to in-hospital cardiac arrests, after implementing a "system of care," some hospitals have reported survival rates as high as 30 to 40 percent. These are just some examples of the unfortunate variation in the quality of care that can be observed. Many who work in the health-care industry are aware of this zip-code lottery of care, which of course impacts the care that people receive with other conditions too; but when it comes to cardiac arrest care, the problem is absolutely endemic.

In the 2012 U.S. presidential primaries, I heard one of the candidates say that if he were elected, he envisioned setting up the first permanent space station on the moon—a feat that would highlight the greatness of the United States. While undoubtedly fascinating, a more "down to earth" feat of greatness for any nation would be to strive to be the "first" in ensuring that its citizens receive optimal standards of care through the complete implementation of the whole of resuscitation science, which as we shall see later in this book has so far been largely lacking. As a result, hundreds of thousands more people could survive a cardiac arrest, and perhaps more important, not suffer permanent brain damage. That would be a truly giant leap for all mankind.

CHAPTER 3

The Formula of Life

LEGEND HAS IT THAT in the early 1400s, a French chemist named Nicolas Flamel discovered the elixir of life. The elixir was reputed to be a substance that could prolong human life for hundreds of years. Flamel lived with his wife, Perenelle, in what is now the oldest stone house in Paris. Located on a narrow street on the right bank of the Seine, the gray stone house was built by Flamel himself and served as his main laboratory. He worked late at night by candlelight, mixing exotic concoctions in an effort to create the elusive potion to sustain life.

Flamel was pursuing an ancient ritual. The search for an elixir of life dates back millennia. According to the Hindu scriptures, the elixir was churned from the deepest ocean waters and called amrita. Throughout the centuries, alchemists searched for the elixir of life to stave off death. They were certain there was a magical potion that could make people live longer, or more hopefully, forever. In ancient China, the emperors were convinced that the right combination of metallic compounds had the ability to effect eternal life. They or-

dered their surrogates to mix cocktails of precious substances, such
as jade and cinnabar, and combine lead with mercury, and even
with arsenic. But more often than not, when the emperors ingested
these concoctions, not only did they not prolong their lives, they
ended up inadvertently poisoning themselves to death!

Flamel took a more measured approach. He toiled for years on
the elixir, which, as legend holds, he created from the philosopher's
stone. The stone was reputed to be able to turn lead into gold and, if
ingested, the resulting substance would extend life for hundreds of
years. In J. K. Rowling's blockbuster novel *Harry Potter and the Sor-
cerer's Stone,* Flamel appears as the creator of the stone that is central
to the plot. In the book, his elixir has clearly worked, as he lives to
the age of 665, while his wife lives to be 658. In fact, Flamel report-
edly died at eighty-eight in 1418, and his tombstone, in which he
carved arcane alchemical signs and symbols, is now on display at the
Musée de Cluny in Paris. Clearly, he did not find the elixir of life.

So is there an elixir of life, and is it possible to overcome death?
Why can't we just live forever or at least for, say, 150 or 200 years?

In the normal course of life, as people age their cells accumulate toxins such as lipofuscin that cause their skin and their organs
to slowly deteriorate over time. So when people get up in years, to
around ninety, their cells stop working properly because they have
built up so many toxins. Environmental factors also contribute to
the cells' aging, as do smoking and heavy drinking. People who
are lifelong smokers or abusers of alcohol often appear to be older
because these vices leave behind toxins that cause premature aging.

But throughout history most people don't die of "old age." They
die "prematurely" as the result of an illness or accident. There are
thus two processes that cause people to die—naturally through
aging, or prematurely through an accident or disease such as a heart
attack or stroke that interrupts the body's functions. Though theo-
retically a person could live far longer than 100 or 150 years, the
reality is that most people do not because as we age our cells pro-

gressively accumulate toxins and this ultimately ends up shutting them down. Hence death through natural causes is due to a biochemical and metabolic process in the cells and consequently the organs. But either way, death is not something mystical; it arises because of the cessation of function in the cells and organs in the body and eventually cellular death.

Now if we could prevent the cells from aging, then in theory, we could prolong life. Scientists and doctors of all stripes are working in this field. One who has garnered attention is Aubrey de Grey, a larger-than-life figure who has polarized opinions and stimulated debate. He studies the biological aspects of aging and contends he has discovered a process that will prevent the cells from building up those harmful toxins. He claims that a tissue-repair strategy he is developing, called Strategies for Engineered Negligible Senescence (SENS), has the potential to allow a very long life span. He has identified seven different areas of cellular decay that occur through metabolic processes and believes that if they are combated through SENS, human life can be prolonged far beyond anything we imagine. Of course, whether or not this will actually work has not been proved, and people continue to die of old age in their eighties and nineties every day.

Although the work that de Grey and others in his field are doing on extending cell life may someday help prolong the lives of people who would die of old age, it will not help those people who die because something breaks down in the body due to an illness, such as an infection, a stroke, or a heart attack. This is the job of resuscitation science.

THE HUMAN BODY IS an incredibly complex machine composed of different components, each with a specific role that enables this machine to function. Like all machines, the body needs fuel to produce energy, which it uses to sustain its biological activities, also known as metabolism. This fuel comes in the form of the foods we eat, but the food alone does not sustain us. Once consumed, these raw products must be burned and converted into energy. This combustion process requires a constant supply of oxygen, the same way an engine needs a mixture of oxygen and fuel to combust. The fuel in the engine is supplied through a gas tank while the oxygen is injected through vents, and in both cases, when they meet, combustion takes place. This creates the energy that enables the other components, such as the wheels, power steering, and lights, to function, but the combustion process also produces waste products. In a car, the waste products are pushed out through the exhaust system; in the body, the gastrointestinal track functions like the gas tank, taking in food that comes from the stomach, while oxygen is supplied through the lungs acting as massive vents. This entire process in both the car and our body is stimulated by the intake of oxygen. Thus, like a car, if our activity revs up as we run, we need far higher levels of oxygen in order to burn more fuel and keep the cells and organs working.

This is the formula of life: taking in oxygen and delivering that oxygen to all the cells in the body in a way that is so extraordinarily precise it makes even the highest-performing supercomputers in cars look basic. This regulation system constantly and effortlessly receives feedback from all organs regarding exactly how much oxygen is required on a millisecond-by-millisecond basis and adjusts the delivery of oxygen to match their requirements so that the cells in each organ can burn the fuel that is in the form of glucose (derived from the nutrients we eat and stored in the body as glycogen) and generate energy. This energy then drives the complex machinery of all the trillions of cells in the human body. However, we cannot store oxygen in the tissues, which is why we can only sprint for a short time or hold our breath for a few minutes. The reason we don't store large amounts of oxygen is that oxygen itself is potentially toxic to the cells (in excess it turns into a strong oxidizing agent such as hydrogen peroxide). Therefore, we must inhale oxygen continuously to feed the cells and keep them working by providing just the right amount.

The cells run much like a highly productive factory that produces special protein-based chemical products such as hormones that are either used locally or sent all over the body (through the bloodstream) to be used by distant organs where they are needed. This is in fact what the enormous amounts of energy provided through the burning of glucose using oxygen is needed for. Just like factories, the cells also have multiple smaller component parts, called organelles, which work cohesively together to produce all the products the trillions of cells in the body need to keep it alive and functioning. Every cell has a membrane on the outside that regulates what goes in and out, much like a wall or perimeter, and specialized small pumps on the wall that can more actively pump substances in and out of the cell. The pumps in the cell membrane take in what the cell needs—glucose that will be converted to energy (as well as many other materials)—and throws out what it doesn't need, such as toxins that can poison the cell. To break down the glucose, the cells rely on oxygen. Small component parts of the cell called mitochondria generate energy by using oxygen to burn ATP (adenosine triphosphate) molecules, which is how glucose is stored in the cells. ATP is thus stored-energy molecules in cells. So the cells need a continuous supply of energy, which they get from burning ATP using oxygen. If we don't give the cells oxygen, the supply of energy runs out and the cells don't function.

When we inhale oxygen through our lungs, it is almost insoluble in our blood, meaning that it won't dissolve in our bloodstream (which could pose a major problem since we need enormous amounts of oxygen every second). This is why we have red blood cells. Red blood cells are like special transporters that bind and carry lots and lots of oxygen through the bloodstream and to all the body's organs. Understanding the formula of life (a phrase coined and often used by one of my old colleagues and mentor, Dr. David Berlin, at

Weill Cornell Medical Center) is simply understanding how oxygen is taken in, or inhaled, through the lungs and then attached to red blood cells that can carry an enormous amount of oxygen through the arteries in the body. For instance, an average man can take in up to around three liters of oxygen through his lungs every minute. A man who is a highly trained athlete can take in up to more than five liters (or well over a gallon) of oxygen a minute, which then attaches to the red blood cells and is transported to the farthest parts of the body. When the red blood cells reach an area where there is a need for oxygen, they release their own oxygen supply and take carbon dioxide that has been produced as a waste product from the activity of the cells and carry it away. The carbon dioxide attaches in place of the oxygen to the red blood cells (the transporters) and is taken back up to the lungs, where it is exhaled. Thus the lungs act like a vent that takes in oxygen when we breathe in and an exhaust pipe when we breathe out. Because carbon dioxide is a waste product, it will become poisonous to the cells if it is not removed, just like the smoke in the factory or exhaust fumes in a car.

The amount of oxygen that is attached to the red blood cells (as well as a tiny amount that actually does manage to dissolve in the bloodstream) is referred to as the oxygen content in the blood, but of course, in order for it to be useful to the body, it has to reach the organs where it is needed. So even if a person takes in a million gallons of oxygen and it is contained in the blood, if the oxygen can't be delivered to the person's organs, the oxygen is useless as far as the body is concerned. To deliver it, a pump is required—and that pump is the heart. Oxygen delivery is a factor that is made up of how much oxygen is contained in the blood (the oxygen content) multiplied by the cardiac output (the pumping of the heart). If for any reason the heart does not pump enough oxygen to meet the body's requirements, a person cannot stay alive and will eventually die. This is why the heart is so vital and why heart disease can kill someone. Also, if we lose too many red blood cells too quickly, such as when we hemorrhage after a car accident or a bullet wound, we will die because without the red blood cells to transport oxygen, we will not be able to meet the body's oxygen requirements. The same would be true if our lungs stop functioning and cannot deliver the oxygen needed. Interestingly, in the event that the body develops a major illness (say following an overwhelming infection) and requires much higher levels of oxygen than can normally be delivered to the organs, then the organs will stop working, resulting in organ failure and eventually death. When we suffer failure of the heart (which itself needs a constant supply of oxygen), we die since the heart is the pump that delivers oxygen to all the organs. So there is a magical ingredient that we need and can help us stay alive. It is not nickel, mercury, or arsenic. It is good old oxygen.

In our normal lives, as we drive to work or sit on the couch and watch TV, the body maintains a constant balance between oxygen requirements and oxygen supply. The red blood cells can carry an enormous amount of oxygen in the blood to not only satisfy daily requirements but also to meet higher requirements. The fact is, in many circumstances we need much higher than normal levels of oxygen. When we become critically ill, say with an overwhelming infection that could kill us, it is often a catch-22: not only can we not deliver enough oxygen, but the cells also need more. This is because the cells have gone into hyperdrive to produce chemicals that will fight the infection. For example, the body might suddenly multiply our white blood cells to try to fight off the infection. The white blood cells act like soldiers and defend against infections from bacteria and viruses. Many chemicals are mass-produced by these cells and released on bacteria to kill them—in a way that is not dissimilar to chemical warfare—and that's why we get a fever.

Think of it like that factory working overtime, because in wartime a country needs many more supplies than in peacetime. But then consider that the country also has problems receiving the raw materials it needs to make those supplies it needs to fight the disease, which is the enemy. This is analogous to the state that Great Britain found itself in during World War II, and why many people believed it was on the brink of collapse. In the body, under these extreme circumstances, we need to be able to deliver much higher levels of oxygen. However, if the body is diseased, it is no longer able to do so, and this can hasten the process of dysoxia and anaerobic metabolism, which eventually causes the organs to stop functioning.

So while we need a constant supply of oxygen, we also need red blood cells to carry it. That is why people die if they bleed heavily due to an accident and become critically anemic very rapidly. Other conditions will also lead to death if the heart cannot pump forcefully and there is not enough blood pressure to deliver the oxygen and the red blood cells through the bloodstream to the organs. When cells reach a point where they do not have enough oxygen being delivered to them, this is called shock, which is a very dangerous state.

Shock is a medical term for the point at which the delivery of oxygen to the cells is insufficient to meet their requirement for oxygen; therefore, the cells can no longer perform their function. The mitochondria that generate heat and energy by using oxygen to break down ATP can no longer break down the ATP molecules, so suddenly the energy source is drained. This causes the cells to build up toxins like lactic acid (which causes cramps in runners). The body has a clever system to try to buy itself some time and prevent the cells from completely shutting down, sort of like a gasoline-powered generator powering a factory that loses electricity. The body starts to burn that lactic acid (through a process called anaerobic metabolism), which will generate some energy; however, it's not enough. Anaerobic metabolism can sustain the cells for a relatively short period of time, but if this period of burning lactic acid without enough oxygen continues, then all the machinery of the cells will stop working. Unless oxygen delivery is restarted soon, then the cells will stop functioning—like the factory relying on a generator that is running out of gasoline, awaiting the electricity to be turned back on.

So it is clear that all the different causes of death share a final pathway and culminate in a state of medical shock. This is characterized by a lack of oxygen being delivered to the organs and, if not corrected in a timely manner, will cause the heart to stop, thus leading to a state that is medically referred to as a cardiac arrest, which is synonymous with death. Causes such as overwhelming bleeding, cancer, infection, poisoning, and heart attacks all can lead to the state of medical shock that, if not stopped quickly, will ultimately lead organs to stop functioning. When this affects the kidneys, the kidneys stop functioning and the person stops producing urine. When it affects the brain, the brain stops functioning, the person loses electrical activity in the brain within about ten seconds, as well as life-maintaining brain stem reflexes that keep us alive by ensuring we breathe and our heart beats, and the person goes into a deep coma. Unlike the deterioration of the kidneys or the liver, which won't kill us immediately even if these organs stop functioning, when lack of oxygen delivery affects the heart and the heart stops pumping oxygen and nutrients around the body, we die immediately. This usually takes a few seconds after the heart stops. So the definition of death is when a person has no heartbeat and no respiration (because the lungs have stopped working due to a lack of oxygen delivery) and is absent reflexes in the base of the brain (brain stem), indicating the brain has also stopped working due to a lack of oxygen delivery. At this point, a person's pupils do not respond to light and he or she develops what is referred to as fixed dilated pupils—hence the reason doctors shine a light on pupils to determine if someone is alive. Cardiac arrest is synonymous with death because the heart is the pump and without the pump, the body cannot deliver the blood that contains the oxygen.

So rather than being a mystical process or a philosophical process, as death is often regarded, it is actually a physical and biological process. Without the body taking in oxygen and the heart pumping oxygenated blood to the tissues, death will occur.

WHAT EXACTLY HAPPENS TO the body after death has taken place? I know that many people would simply say: "Well, that's it. It's the end." But is it really? Let's imagine for a moment that we could step into the body after death and witness what takes place. We know that the heart stops pumping the blood carrying oxygen, that elixir of life. Therefore, the cells no longer have oxygen entering them. Without oxygen, their pumps will shut down, and no energy will be produced. But what is actually happening is that we are entering a second phase that is the period "after death"—the stage that corresponds to the gradual death of the cells, which only starts after we have died and takes many hours.

It is likely that this concept is unfamiliar to most people. To illustrate what happens to the cells after death when oxygen is no longer being delivered to the cells, consider the scenario that occurred in New Orleans during Hurricane Katrina and what precautions the city has taken to prevent such devastation from happening again. There are many individual homes in the different wards of the city. To keep people alive in those houses, a certain temperature and oxygen level must be maintained. The houses also must have water but not too much water. To protect those houses, the city also has protective levees, just as the brain has a protective mechanism for its cells that regulates blood flow into it, because too little or too much blood flow can be highly damaging. During a crisis, much like the levees that regulate water levels, the brain (in a process called autoregulation) regulates the amount of blood flowing into it. This prevents a sudden increase and especially a sudden drop in blood pressure from disrupting blood flow and hence delivery of oxygen to the brain.

This works wonderfully except in extreme cases when something (such as an infection or bleeding) causes the blood pressure to become too critically low. Under these conditions adequate oxygen isn't being delivered to the brain. When oxygen delivery to the brain drops below a critical threshold level, at first a person may become agitated and then go into a coma. If the blood pressure drops even more, then the brain completely stops working within seconds. The constant flicker of electrical activity that is the hallmark of an active functioning brain, like city lights flickering away when we look down at them in the night sky, simply stops. Everything turns into dead silence—the brain is overcome by a "flatline" state. We can measure this electrical activity (or lack thereof) from the surface of the brain using a machine called an electroencephalogram (EEG). Of course, the most extreme condition that stops blood flow to the brain is a cardiac arrest, or death, resulting in no blood pressure at all since there is no heartbeat. When autoregulation of blood flow and hence oxygen delivery to the brain fails, the cells first go into a panic mode and then respond by going into a toxic fury that begins within minutes.

Now imagine that each cell is analogous to each home and also has its own pump built into the wall to regulate what goes in and out. The brain cells need optimal conditions to generate electricity. Electricity is created when there is a large gradient of sodium and other substances such as calcium (with high levels staying mainly outside and a little amount inside the cells) to generate electricity. It is the movement of mainly sodium and other chemicals such as potassium and calcium in and out of the cells that leads to electricity. If we have too much calcium in the cells, they fail. These pumps need a constant energy supply to work, and after a person dies, because there is insufficient oxygen and ATP stored up, the cells start to go through their own process of death.

This destruction and devastation take place sequentially during a process, and so from a medical perspective, if we identify and understand the individual components that take place after the devastation of death has set in, we can try to slow down and stop those changes that are taking place. Then we can also stabilize, restore, and finally repair and return the cells to a working order. This is, in a nutshell, the science and art of resuscitation. This is exactly what we have learned we can now do with respect to the brain and the rest of the body after the devastation of death and cardiac arrest when there is no oxygen being supplied. It is not simple by any means, but with a good system of care it can be accomplished.

What happens in a person who dies is much the same as a city flooding. A person has billions of cells. These cells function best with a small amount of calcium—the same way a house needs a controlled supply of water. To maintain the balance, pumps are needed to remove the excess calcium. Because oxygen cannot be stored in the cells, as soon as oxygen stops, a downhill process begins. Once the heart stops pumping blood, which contains the oxygen, within four minutes the body will consume all the stored oxygen and energy and the pumps will begin to fail.

During this process, there is a massive release of various toxins from inside the cells as they start to swell up and the membranes, which act like walls around the cells, become damaged. All the while, the cells become more and more acidic, swollen, inflamed, and damaged. The pumps that normally regulate what comes in and out of the cells don't work, and calcium starts to flood in from outside, causing the cells to swell even more and therefore damaging the walls of the cells' membranes. The cells' membranes begin to crack holes inside of them, and more and more calcium floods in. This vicious cycle causes the cells to become more and more damaged. The cells experience calcium accumulation that eventually leads to toxicity (called excitotoxicity). They will eventually go through a biochemical program called apoptosis that causes them to quietly shrivel up internally or actually rupture outwardly through a process called necrosis. These processes by which cells eventually die take place through a chemical chain reaction that requires the activity of chemical catalysts called enzymes. Therefore, one way to combat the damaging effects of death, and to halt and reverse it even after it has set in, is to slow down these chemical reactions by targeting and blocking the activity of these chemical catalysts in the brain and other organs. Without the actions of the enzymes in brain cells and cells in other organs, they can't "die" as quickly because even cellular death is a chemical process. If you stop the chemical process, you halt or at least slow down cell death.

This is also exactly what happens during a common stroke. In a stroke, the cells in a limited area of the brain are deprived of oxygen, and they go through this entire process of death. That's why some-one with a stroke can have severe disabilities. Those cells affected by a lack of oxygen delivery have died over the course of a few hours. However, medically speaking, death is a global stroke in which the entire brain is being deprived of oxygen. But in the same way we can limit and reverse brain cell damage from a stroke if we can treat the patient quickly (for a common stroke affecting the brain, most people now quote a time frame of approximately four and a half hours), we can also reverse a global stroke—and therefore reverse death.

During either of these processes, cells damaged by a lack of blood flow and the delivery of essential nutrients and oxygen no longer function; yet they have not been fully destroyed. Thus, if blood flow and the delivery of oxygen and essential nutrients are restored during this time, the cells may either partially or fully recover. So in order for us to fight death, we must not only stop the deterioration process but also prevent it from reaching that point of no return in a controlled manner. Even though cells are going through rapid dete-rioration during this process, these cells that make up the brain and the other organs are still viable. Again, they are not functioning, because they are all going through this debilitating change, but they can be rescued, built back up, and made to work again. However, if we leave them for several hours (as in someone who has had a stroke many hours earlier), they will become completely shriveled up and die. This is the medical opportunity and challenge we have been offered through scientific progress today—the ability to go beyond death and come back safely.

These methods are central to resuscitation science. This is illustrated by the incredible understanding that cells can be taken from a dead person's brain many hours after a person has died and been taken to the mortuary and then grown in the laboratory. In May 2001, scientists at the Salk Institute in La Jolla, California, reported that they were able to take brain cells from cadavers and grow them in the laboratory. The scientists showed that these cells taken from people who had been dead for a number of hours could not only grow but also divide and form specialized classes of brain cells. The scientists' work was focused on growing brain stem cells called neuronal progenitor cells, but it clearly shows the viability of brain cells many hours after the blood and oxygen supply has been cut off, after the heart has stopped beating, and after death. "I find it remarkable that we all have pockets of cells in our brains that can grow and differentiate throughout our lives and even after death," said Fred Gage, a professor at the Salk Institute and senior author of the study.

As a result of such pioneering work, most remarkable of all, brain cells can now be taken from people even four hours after death and grown in the laboratory—meaning they are still viable. In fact, at a recent conference in New York copresented by the Nour Foundation—a nonprofit and NGO that explores meaning and commonality in human experience—and the New York Academy of Sciences, I met a scientist who told me that she routinely harvests brain cells from cadavers and grows them in the laboratory for research purposes (the people she studies had consented to their cells being removed after death). Even four hours after death, she can take a biopsy of brain and grow it in the laboratory.

Estimates vary on how long cells can survive without a blood supply in different organs (and therefore without oxygen after death) depending on factors such as the type of tissue involved and the ambient temperature. Bone is the most tolerant at up to four days. Skin can survive up to twenty-four hours, and fat up to thirteen hours. Nerve cells and brain tissue (neurons) are thought to remain reversible for up to eight hours. Even with the variations, what this tells us is that there is a significant period of time after death that cells can be brought back.

ONE OF THE MAJOR discoveries in the last ten years that has allowed us to halt the chemical reactions that take place in the cells of the body and brain after death has been the finding about the importance of cooling the cells down. A chain reaction occurs in the brain with all these chemical reactions that are going on, which, like any chemical reaction, depends on temperature. Heat speeds up chemical reactions; cold slows them down. Cooling cells deprived of oxygen slows down the activity of the enzymes that regulate chemical reactions in cells and hence reduces the harmful processes that take place after the toxic fury has set in and gives us time to restart oxygen supply to the cells.

In early 2011, I received a call from a colleague in England telling me an incredible story that illustrates this point. Medically, the story was about slowing down the deterioration process of cells deprived of oxygen so that a man who had been dead for three and a half hours returned to a normal life. Though this man had no heartbeat and was not breathing for three and a half hours, he survived and, more important, was discharged home from the hospital without brain damage even though others in his situation don't survive. How was he able to regain functionality after his heart had stopped for so long? Specifically, what kept his cells from becoming completely flooded out and dying irreversibly?

The story says much about cell viability when the cells are cooled. The man, a fifty-three-year-old named Arun Bhasin, was walking home from a party in East London in 10° Celsius (50°F) weather when he collapsed. A passerby found him suffering from hypothermia and called paramedics. When the man arrived at the hospital, he was in bad shape. His body temperature had dropped to 30°C (86°F)—37°C (98.6°F) is normal. As doctors began treating him, he suffered a cardiac arrest.

However, because his body was already cooled, the metabolic activity in the cells had slowed down. He also had the luck of the draw on his side as to the technology that was available at that hospital. Two experts in resuscitation medicine, Dr. Nigel Raghuntah and Russell Metcalfe-Smith, were on duty. They hooked him up to a machine called the ZOLL AutoPulse, which automatically administers consistent and very high quality chest compressions, while they began looking for the cause of cardiac arrest. Three and a half hours later, his heartbeat was restarted.

Not only did Arun Bhasin live, he was able to regain full functionality with no cognitive impairments—because his body was cold while his heart was not beating and his cells were not getting oxygen. When he arrived at the hospital, his temperature was two degrees below the optimal cooling level of 32°C (90°F) for that situation. Studies show that for every one degree Celsius that we bring down body temperature, we reduce brain metabolic activity by about 6 percent. So if we can cool the body from 37°C to 32°C (98.6°F to 90°F), we can reduce metabolic activity and thus the rate by which brain cells go through their own death by roughly one-third. That means the brain cells, even though not functioning, will be less likely to become permanently damaged while the heart is stopped and not providing them with oxygen. Although Bhasin was slightly colder than recommended, the fact that he was cold functioned like a braking mechanism on cell deterioration. It didn't allow the calcium to rapidly back up and flood the cells with toxins. It slowed the process of apoptosis and necrosis that is the hallmark of cell death, because it prevented the enzymes from working and thus prevented the chemical reactions needed for cells to die.

Most of the news reports of Bhasin's story talked about what a miracle it was that he was both alive and not brain damaged. Well, it wasn't really a miracle; it was resuscitation science at work. His case is part and parcel of understanding what causes brain and other vital organ damage and how it can be stopped. His brain cells were basically kept cold so all the processes that lead to cell death were slowed down significantly, thus giving his doctors time to try and restart the heart and reverse the destructive damage that ensues after the heart has stopped. That's why we put food in the fridge; it stops the decaying process. Though cell deterioration isn't bacterial, cooling the cells slows the enzymes in the cells that generate the chemical reactions.

Bhasin also had another advantage of modern technology. The hospital where he suffered the cardiac arrest used the ZOLL Auto-Pulse chest compression machine. This machine gives standardized, consistent chest compressions, thereby eliminating human error and fatigue. Again, as discussed earlier, today it's a complete zip-code lottery if we end up in a hospital that uses such an automated machine or has a truly state-of-the-art system to manage cardiac arrest, because many hospitals don't.

The advances in resuscitation science are new and continually developing, and they can be overwhelming to grasp. In many ways, they can overtake our social psyche that is based on the idea that there is a physical moment of death. We have inherited this firmly held notion about life and death through many millennia; it has been passed on from the classical Greeks and lived through the Renaissance and Victorian times and remained unyielding until now. Even the change that has come about since then up to this day has been very slowly understood. Much of what we accept as real is socially and not scientifically determined. But we have now moved into this gray zone that corresponds with a period "after death," which has come about as a consequence of humankind's persistence in attempting to reverse death after it has set in and which, as we shall see, has significant ramifications for us all.

CHAPTER 4

Reversing Death

Dᴜʀɪɴɢ ᴛʜᴇ Dᴜᴛᴄʜ Gᴏʟᴅᴇɴ Age in the seventeenth century, the beauty of Amsterdam was created when city planners supervised the digging of hundreds of miles of canals. The three main canals formed concentric belts around the city, and they were bordered by markets, restaurants, and row houses. The romance of the canals was highlighted by narrow pathways that allowed people to walk close to the water, particularly along the stately area known as the Golden Bend that housed the city's great mansions. However, by the mid-eighteenth century, immigration to cities had pushed up Amsterdam's population and created a unique problem along the winding canals: a high number of people would fall into the canals and drown.

Because there was no foolproof way to prevent people from falling into the canals, people began looking at ways to save those who did. At the time, it was believed that if someone's lungs filled with water due to drowning, air could be blown into the lungs using a fireplace bellows, which in effect turned into something of a precursor to the modern ventilator. So in 1767 when the Amsterdam Rescue Society was formed to solve the drowning problem, one of its solutions was to place a fireplace bellows at intervals along the city's canals; therefore, if someone encountered a drowning victim, they could pull the person from the canal and use the fireplace bellows to resuscitate him or her. It was similar to how we today place defibrillators in airports and other public places to aid people who have a cardiac arrest and die. Within four years, the society claimed to have saved more than 150 people.

The fireplace bellows method came about during the Enlightenment. It was a time when scientific understanding paved the way for people to begin attempting to prolong life by trying to somehow reverse death. Though life and death had long been thought of as opposites and it was believed that only God or one of his prophets could intervene, people had now begun to attempt to control their own destinies by using medicine rather than leaving it to higher powers. As odd as the fireplace bellows method sounds, it was actually quite innovative compared to the ancient efforts used to attempt to resuscitate the dead. For instance, Galen of Pergamon, the legendary Greek physician, associated life with warmth, so people in his time did anything they could to warm up the dead and hence "revive" them. They covered the dead in hot water, warm ash, or heated excrement. In some parts of the world, people believed that stimulating the corpse was the answer, so they blew smoke into a dead person's anus, an equally undignified, unsuccessful maneuver, while others tickled the throat with a feather. These desperate, almost entirely ineffective procedures ruled the day until the Enlightenment.

Because so many people who died prematurely in European cities died of drowning, the major focus remained on ventilation and trying to get air into the lungs. For those who died on land, the rescue societies that had sprung up across Europe found a different technique. They did something that sounds odd but was actually very innovative. They put the dead person on a barrel and rolled him, or they laid him across a horse on his chest so he would bounce up and down when the horse was set into a trot. Though these were basically commonsense treatments, people were clearly beginning to realize that chest movement had something to do with reviving people. However, these arcane ventilation and chest compressions did not work on their own because the rescue societies did not understand that stimulating the heart to restart was critical to resuscitation.

From 1767 to 1949, the major focus in resuscitation remained on ventilation. In 1858, British doctor H. R. Silvester took these ideas one step further. In his article "A New Method of Resuscitating Still-Born Children, and for Restoring Persons Apparently Drowned or Dead," published in the *British Medical Journal,* he advocated using the patient's own arms as levers to pump the chest in and out, aiding inhalation and exhalation. What quickly became known as the Silvester method was the first time modern CPR techniques incorporating both breathing and chest movement flickered into view. In the early 1900s, cardiac massage was introduced to restart people's hearts that had stopped. To reach the heart, doctors would open the chest of the patient and then directly massage the heart in an effort to restore heartbeat. Though this was successful in some cases, it was done infrequently due to the obvious danger of cutting open a person's chest and rubbing the exposed heart. Still, no one had put together both components, or even come up with a way to create and maintain circulation and also artificially oxygenate the lungs, as had been done in a very primitive manner with the fireplace bellows method all those years ago.

A major breakthrough came in 1949 when James Elam, an anesthesiologist, introduced mouth-to-mouth breathing as a way to oxygenate the lungs. Elam had read historical accounts of newborns who had trouble breathing being resuscitated with mouth-to-mouth breathing. Elam was working in a Minneapolis hospital in the middle of a polio outbreak. As he was walking through one of the wards, a young boy who had turned blue and wasn't breathing was being rushed into an attending room. Instinctively, Elam grabbed the boy and inflated his lungs. The boy's color returned and so did his breathing. Elam then set about studying mouth-to-mouth breathing as a means of reoxygenating the lungs, and he was soon joined by scientist Peter Safar. After Elam and Safar conducted a series of experiments, the technique of mouth-to-mouth ventilation as a way to oxygenate lungs was adopted by the U.S. military in 1957 and endorsed by the American Medical Association in 1958.

It was not until the early 1960s that scientists and doctors put the pieces together and discovered that closed chest compressions integrated with mouth-to-mouth breathing could revive a person who had stopped breathing and had no heartbeat. There were two elements to this, the artificial circulation of air and the artificial circulation of blood, as a person with a stopped heart would not be able to circulate oxygenated blood. Elam and Safar had shown that oxygen could be artificially circulated in a person's lungs with no heartbeat (and therefore no lung function) through a respiratory tube. The artificial circulation of blood in a person whose heart had stopped came into view as a result of an experiment performed on anesthetized dogs in a laboratory at Johns Hopkins University by William Kouwenhoven and Guy Knickerbocker. They discovered that just by physically pushing electric paddles into the chest they could cause blood to circulate despite the fact that the heart was not pumping. This force was actually strong enough to create a pulse in dogs where there was none and thus was a means to circulate the blood. Along with James Jude, a surgical resident, Kouwenhoven and Knickerbocker then found that using their hands for chest compressions with the body lying on its back also achieved the same result. They had found a way to squeeze blood out of a stopped heart and for the first time create an artificial circulation. This experimental technique was soon tested on a woman who suffered a cardiac arrest and died in a hospital. Consequently she had no blood pressure and no pulse. Normally, the only thing they could have done would have been to cut open her chest and physically massage the heart with their hands, but they were not in an operating room so that was impossible. Instead they applied this new system of external chest compressions and performed mouth-to-mouth breathing; successfully and incredibly, they managed to restart her heart and get her breathing again.

These two techniques—the artificial circulation of air and the artificial circulation of blood—were further developed and combined with providing an electrical shock to the heart in order to restart it, and all three components were merged to create what became the resuscitation technique we now know as advanced cardiopulmonary resuscitation. This was presented to the medical community by Jude, Knickerbocker, and Safar on a world speaking tour in 1960–61, and later put in a medical training video made in 1962 entitled *The Pulse of Life*. In 1963, the American Heart Association endorsed this revolutionary technique of CPR as a method to restart the heart of a person who had suffered a cardiac arrest and died. For the first time, we had found a way to routinely reverse death.

However, although we take CPR for granted now, like many new discoveries in medicine and science, for almost the first ten years many people were incredibly skeptical of these new methods. This is not dissimilar to what we often observe with anything new in science until eventually things become accepted by enough people who are opinion leaders and then gain broad public acceptance and are embraced as science. This dovetails with what we discuss later regarding the way that reality is actually defined and developed by society.

Throughout the late 1960s and into the 1970s, there was a greater dissemination of information on CPR, such that by the mid to late 1970s it had become widely accepted as an effective lifesaving method. It also became clear to doctors that keeping people alive required a chain of components and if one chain of that component was missing, the patient would not survive. This early chain resulted in four components that converged to create modern cardiopulmonary resuscitation—chest compressions, ventilation, electricity, and the use of certain potent drugs called vasopressors to help create a blood pressure by squeezing the blood vessels inside the body while resuscitation was going on. Essentially, these are the current bread and butter of how doctors today deal with patients in cardiac arrest. To overcome the lack of oxygen delivery to the organs, we try to create a circulation artificially by pressing on the chest, and we try to oxygenate the patient by giving oxygen through the lungs with a ventilator if we can, or if not through mouth to mouth. Then in cases where the heart has an abnormal electrical rhythm, we also administer electricity to shock the heart back into a normal rhythm, and throughout this time we give the patient drugs to regulate the blood pressure by squeezing the person's blood vessels.

Putting these into practice came with the creation of a system of care that is now called Basic Life Support (or BLS), which is the sequential mouth to mouth with chest compressions that is still taught everywhere. In hospitals, where there is more expertise and specialized skills available to treat patients, the technique was taken further, resulting in a process that is now known as Advanced Cardiac Life Support (or ACLS). ACLS is an extension of BLS. After BLS has been performed, an endotracheal tube is inserted to artificially breathe for the patient. Simultaneously, concentrated doses of the body's own hormones such as adrenaline or vasopressin are administered to squeeze the blood vessels and elevate the blood pressure to some extent. In cases where the heart has completely stopped because it has gone out of rhythm, a condition called ventricular fibrillation (the heart has electrical activity, but it is so dysrhythmic that it is not conducive to the heart pumping), we can shock the heart using a defibrillator in an attempt to resume normal electrical rhythm, which will cause the heart to beat regularly. Through starting BLS and then leading on to ACLS, doctors found that death could be reversed.

By the 1980s, although there were more and more miraculous reports of people being saved and being pulled back from the clutches of death, there were also many stories of people succumbing to death despite modern resuscitation. For every one person saved, a multitude of people didn't survive. Clearly, things weren't as wonderful and rosy as everyone had hoped. A major concern was beginning to gnaw away relentlessly at doctors' minds and eat away at their consciences. Yes, while effective CPR—particularly if practiced with all the expertise and aplomb prescribed in ACLS involving repeated cycles of chest compressions, ventilation, defibrillation, and drugs—worked and enabled doctors, paramedics, and nurses to bring people back even after they had started their descent into the abyss of death, it was also clear that if resuscitation was prolonged and went beyond ten to twenty minutes, then the chances of success were remote. Furthermore, if resuscitation was started late and a person was left essentially dead for too long (i.e., without a heartbeat for more than about ten to twenty minutes before CPR was started), the chances of success were equally slim. Another problem was that although doctors had become pretty effective at restarting the heart (in actual fact some could restart the heart in half of all the people they worked on), after all that hard work, more than two-thirds of those initially saved would die again. Consistently, many people who had died once and been revived would die a second or third time in the subsequent hours to days after being resuscitated, and when that happened, they could no longer be revived. Powerless and helpless, all doctors could do was watch and wonder why.

The statistics were really bad. Although in hospitals up to 50 percent of those whose hearts had stopped could be revived, everyone knew that the sad truth was that only about 15 percent would actually leave the hospital alive. If a person was unfortunate enough for his or her heart to stop outside of a hospital setting, say at home, the odds of survival were even worse. At best, about 5 percent of these people would survive to a point where they could be discharged home; a survey of many major cities in the United States showed a survival rate of 1 to 2 percent, and in some places it was literally zero. Of those who had miraculously survived, like soldiers limping off a battlefield, many would end up with the catastrophic consequences of disability and permanent brain damage, or even more horrendously, the dreaded final consequence of having survived a period of whole brain stroke (better known by its technical name of anoxic brain injury) that everyone feared the most: a persistent vegetative state. If left unchecked, for many this was the ultimately debilitating result of the fire left by the toxic fury that had taken place in the brain cells after the heart stopped and was restarted. Furthermore, up to 50 percent of survivors reported memory defects, depression, and even posttraumatic stress disorder. These were clearly terrible statistics, and these outcomes would turn any euphoria felt by doctors at the initial success of starting the heart into utter despair. Whereas in the old days people would just die and would never make it out of death, now there was the added burden of people being revived only to be left to live a life of permanent physical and mental scarring.

By analogy, it was like a group of scientists realizing that, after the initial excitement of being able to send people into space had subsided, the reality of returning those people safely was difficult. It wasn't pleasant, and it certainly wasn't without peril. Too many people were either burning during the descent, or if they survived the descent, they were forever carrying the physical and mental scars of their journey with them. Thus, while after centuries of trying, serendipity had provided humankind with the means to go beyond and reverse death, it hadn't prepared us for the perilous return journey, and while everyone had largely been focused on the journey into space, they were now all beginning to wonder how they could return safely and alter the depressing survival statistics. The answer was beginning to emerge from an unlikely source behind the Iron Curtain.

In 1972, in the midst of the Cold War between the United States and the Soviet Union and the battle between the countries to top each other in space exploration, a stern, stocky, and quite stoic Soviet scientist named Vladimir A. Negovsky made an incredible breakthrough. In an article published in the scientific journal *Resuscitation,* he wrote: "In the first stages of the development of the science of resuscitation or 'reanimatology' [as he called it], research workers have been limited mainly to the study of the pathology of death, and to the elaboration of a series of techniques of resuscitation. We now have at our disposal some knowledge of the process of disintegration of physiological functions during the dying of an organism, and of their restoration during resuscitation." Negovsky noted that there was much evidence that the organism experienced a specific pathological condition after resuscitation, the condition he called "post-resuscitation disease."

In a nutshell, Negovsky had discovered that surviving death was no more natural than being shot into space and returning to earth. While a person returning from space would burn in the atmosphere without proper protective measures, people who had been revived and brought back from death would also burn from a toxic fury that ensues in the body after being resuscitated. The reason that people were dying was actually a direct result of having been resuscitated in the first place. There were serious consequences to resuscitation, and he had begun to discover them. Remarkably and ironically, it became clear that the system that had saved people was actually killing them again.

Negovsky discovered a unique medical condition that seemed to solely afflict those who had survived being resuscitated. Negovsky's work and others that followed made it clear that the mechanisms underlying this postresuscitation disease involved a two-step process. First, doctors now referred to the state of "whole-body ischemia"—meaning total-body oxygen deprivation brought about by the heart stopping and death itself. Second, there was another phenomenon called reperfusion syndrome. This only happened when the heart had been restarted. It was discovered that when oxygen (which normally keeps cells alive) is introduced back into areas that had been deprived of oxygen, oftentimes it turned into a powerful toxic substance that instead of saving the cells would hasten their death. This was something that Dr. Lance Becker, a prominent emergency physician at the University of Chicago who now works at the University of Pennsylvania, recognized as being an area of major importance. He explains that for so long the issue of oxygen was considered such a simple, urgent truism that it went without question: restoring *full* oxygenation was the first priority. But when he subjected this assumed scientific theory to real scrutiny, what he discovered shocked himself and the medical establishment. Cut off from their oxygen supply, heart and brain cells enter a hibernation phase, fending off their own slow destruction in a kind of preprogrammed effort to stay viable for as long as possible. If full oxygenation is reintroduced at this stage, when the cell is most vulnerable, the cell actually *dies faster.* This was the opposite of what anyone expected, so a lot of subsequent research has focused on how we might extend this hibernation phase and reintroduce oxygen more slowly and safely. It was further discovered that in patients with cardiac arrest, this ongoing damage and killing of the cells all over the body led the cells to release numerous chemicals into the bloodstream, triggering a massive inflammation all over the body that was so severe it resembled the whole-body inflammation that people get who have been afflicted by massive and deadly infections, such as severe bacterial meningitis. This massive inflammation would devastate all the major organs that it came across, causing the lungs to fill up with water, and the liver and the kidneys to fail. As for the brain and the heart, they didn't fare much better. One by one the organs would fail within hours to days, and finally when the heart failed and stopped beating as well, there was no return because the underlying inflammatory response couldn't be stopped.

The relationship of initial whole-body ischemia to the secondary reperfusion syndrome is like the relationship between an earthquake and a tsunami. Much like an earthquake, the initial whole-body ischemia arising from cardiac arrest does kill people, but the real problem for those resuscitated is the secondary massive tsunami that engulfs the person and decimates everything in sight. The magnitude of the tsunami is directly proportional to the magnitude of the earthquake; the bigger the earthquake, the bigger the tsunami.

After we die and recover, various enzymes and chemical processes either become activated or inactivated in the cells in the body that will take us toward permanent cell death. The only exceptions are for those cases where the period of cardiac arrest and death before receiving resuscitation has been very short, or for those cases in which doctors intervene correctly. In this process, cells are hit by the consequences of a prolonged period without oxygen and then hit again by the consequences of having oxygen, and they start their own process of death. They need to be rescued; otherwise, they will continue on their own trajectory of death. Therefore, even if the heartbeat is restarted and circulation restored, the cells will continue to die through a process of chemical reactions involving activation of various enzymes and pathways. These chemical and enzyme pathways are quite elaborate and very complex. They involve activation of enzyme pathways inside the cells that usually lead to death either by a process called apoptosis or necrosis. Apoptosis is a programmed code built into the cells that causes the cells to "commit suicide" if something devastating damages them, resulting in their actually shriveling up and imploding. Necrosis is a process by which the trauma to the cells is so extensive that the cell walls and all the cells' contents literally explode outwardly. Unless something is done to put the brakes on this cell deterioration and reverse the process, the person whose heart has been restarted will die again. So to survive, a person must endure both the earthquake and the tsunami.

THE GOAL IN MEDICINE is generally the cure. Medicine is often interested in the pathophysiology of disease, though not for the sake of it but as a means of identifying the opposite, which is the cure. When studying a disease, in order to figure out how to cure it, we must learn the details of the disease. Basically, researchers need to work backward, looking at the opposite of the goal. Once we learn what that is, then the treatment becomes quite easy; we just do the reverse. But it is imperative to simultaneously understand both aspects, or we will not likely find the cure.

Consider rheumatoid arthritis, a condition that causes the joints in the fingers and other parts of the body to swell. Before the medical community knew anything about the disease, doctors would attempt to solve the problem by putting ice on the knuckles to reduce the swelling, as if someone had hit his or her hand with a hammer. This was unsuccessful because the true problem was not understood. Rheumatoid arthritis is an autoimmune problem in which cells attack the joints and cause inflammation. As a consequence, this can cause inflammation in the organs such as the heart and the liver, resulting in a multisystem disorder. An ice pack on the knuckles is obviously not going to cure this. Today we give medication that both targets the immune system and also suppresses inflammation in the body. But we can only do that because we understand the concepts.

Reversing death requires a similar understanding. As we have seen, multiple stages are involved in death. First, before death has even taken place, we know that a disease or condition either leads to inadequate oxygen being delivered to the cells in the organs or, in the case of poisoning, stops cells from using oxygen, leading to the blood pressure dropping to critically low levels—a process we refer to as shock. If left unchecked, shock will cause the organs to stop working, and when the heart stops, this is death and leads to irreversible cell damage and eventually cellular death. Since we know cells can potentially remain in a state in which they can be saved for many hours after we die, the big unknowns during this process are reflected by the questions: How can we best intervene and prevent cells from being irreversibly damaged in someone who has died? When does our consciousness, or the real self, become lost? And what happens to that consciousness? We don't know the answers, but these questions clearly warrant research into the period after death but before complete and irreversible cell death has set in.

Medically speaking, all the causes of death that we can ever imagine first lead to shock and then ultimately death by one of four physiological processes. The first is when there isn't enough blood in the system. An example of this is severe bleeding from a gunshot wound; even though the heart is pumping fine, there isn't enough blood to pump around. The second is when the heart pump itself fails, as in the case of a massive heart attack. The third process is actually when the blood vessels in the body dilate and open up enormously due to massive inflammation in the body, dangerously altering the mechanics of the circulation. This occurs when an overwhelming infection causes the body to suddenly release inflammatory chemicals that cause all the blood vessels to dilate and engorge, resulting in all the blood in the central portion of the body suddenly being pulled into the arms and legs, thus curtailing circulation. This is also what happens during an anaphylaxis reaction such as with a peanut or shellfish allergy. The fourth occurs when a physical blockage in the major blood vessels prevents circulation of blood from taking place. So, in these situations, the heart pump is working fine, there is enough blood in the system, and the blood vessels that carry the blood are dilating and narrowing appropriately, but there is a mechanical obstruction. For example, a massive blood clot could lodge in one of the major blood vessels and block the entire circulation, or fluid could accumulate around the heart and squeeze and constrict the heart so hard that it won't be able to pump. Regardless of the type of shock, if it is not reversed, then the heart will eventually stop.

But how effective are chest compressions at creating circulation and mimicking the pumping of the actual heart? We know that without chest compressions after a person dies, the blood pressure is zero—meaning there is no circulation. But if chest compressions are done really well—hard and fast at a specific rate and depth—then we can generate a blood pressure that while low is better than nothing. Typically, we may get a blood pressure up to 50/20, maybe as high as 80/30 in some cases if someone is incredibly good at performing chest compressions. But we have to remember that a normal blood pressure is around 120/80, so the blood pressure created with chest compressions, while better than nothing, is still too low.

Giving potent drugs called vasopressors to squeeze the blood vessels also helps push the blood pressure up a little more. In one study by Dr. Mark Angelos, at Ohio State University, and his colleagues, the drug epinephrine (adrenaline) was used on ten people who had CPR performed on them. Their blood pressure with chest compressions alone before the drug was administered was in the range of 47/18. With 1 mg of epinephrine, their blood pressure increased to 69/27, which was better but was still very low. What is also important to realize is that of the two numbers used in blood pressure calculations, the lower number (called the diastolic) is the one that determines how much blood actually gets into the heart and the brain itself, because it is only when the heart pump has stopped squeezing that the blood can flow into the heart itself. The same is also true of the brain. What this shows is that the diastolic pressure doesn't increase much with either chest compressions or with the administration of potent drugs, which is telling us that we are not getting adequate blood into the brain or the heart.

Everybody who has watched TV has seen at one time or another someone perform chest compressions. The most important factor is the quality of the chest compressions, which involves consistent pressure and depth and the proper amount of time between compressions. Numerous studies show that if people receive CPR by the best-trained experts (for example, those people who teach CPR skills to others), they actually only receive adequate chest compressions around 40 percent of the time. Of course, the reality is that most people who work in hospitals or ambulance services are trained by the experts but aren't at the same level as the experts themselves. So we face a major challenge because the quality of CPR delivery is highly variable and many times less than adequate.

One of the most compelling studies in this area was conducted by Dr. Dana Edelson of the University of Chicago. She focused on improving the quality of CPR to improve survival outcomes. This sounds pretty straightforward, but it's actually complicated because generally no feedback system is in place to determine if the CPR being performed in the hospital is reaching the correct depth and quality.

Hypothetically, if you deliver the best-quality chest compressions, studies have shown that you will be able to give back to the patient about 25 to 30 percent of the blood flow the person would need. This is even with the administration of the potent vasopressor medications such as epinephrine because the blood pressure is still too critically low. In reality, if the chest compressions are not perfect, that percentage will be less. We also know that overbreathing kills people, because breathing out normally takes longer than breathing in. The person performing CPR must allow a period of time for air to be exhaled from the patient. The problem is that if you push air into a patient's lungs and don't allow enough time for air to be exhaled before pushing more air into the lungs, a phenomenon called breath stacking occurs.

Breath stacking is what it sounds like. It means that you are stacking breath after breath inside the chest, which causes the pressure inside of the chest to become elevated. The first consequence of the pressure inside the chest cavity becoming high is that it starts to compress all the organs and therefore restricts the blood from getting to the heart (thus leaving the heart critically low of blood). The second consequence is that because you are overbreathing, you are blowing out too much carbon dioxide. Though to some extent you need to do this, you also need a certain amount of carbon dioxide in the circulation because carbon dioxide is one of the things that determines whether the blood vessels that take blood to the brain are wide open or narrowed down and thus regulates the flow of blood to the brain. If you think of two blood vessels being a highway, the carbon dioxide determines the number of lanes, or how open it is. When too much carbon dioxide is blown off, the blood vessels squeeze down and constrict. When they constrict, the blood that was supposed to go to the brain—which is only at best 30 percent during excellent CPR—suddenly becomes much less because the highway is restricted. For a long time, this problem was not recognized because the survival outcomes have been so poor for so many years, but as we have zeroed in on how to improve outcomes, this has become one of the focal points.

In one case I was involved with at one of the hospitals I worked at, I received a call to come to the ER because a fifty-nine-year-old man who had been pronounced dead was in fact alive! The doctors who worked on him had told his wife that he had died, but when they were preparing his corpse, he began breathing. Most likely what had happened is that they had been overbreathing for him (i.e., they hyperventilated him), which caused air and hence the pressure that is exerted by forcing air into the chest to build up, and this had blocked his circulation by squeezing down with pressure on the veins that carry blood back to the heart during CPR. After they had stopped and pronounced him dead, the air had escaped out of the chest and his blood vessels had a chance to recoil; then blood could get back into the heart once more, and it started beating again. If the effects of breath stacking are not recognized, it can itself hasten death and is called "death by hyperventilation" in the medical literature.

I experienced a similar situation with a ten-year-old boy. The boy was being administered anesthesia for an operation and his heart stopped. The doctors frantically gave him CPR, but he was not responsive. Again, they were banging air into his chest so frequently that there was no chance for the air to escape out and for his heart to respond. After forty minutes, they stopped for a short time to connect him to a ventilator in a last-ditch effort to save him. During that brief period, when air wasn't being pushed in so frenetically, the air in his chest escaped out, and his heart rested and restarted.

These two cases illustrate what happens when someone is returned from the dead. They also show how without attention to minute details, our success could be compromised. Much of the problem is that we are dealing with a very stressful situation, but feedback systems are not routinely employed in hospitals to grade and inform doctors how well CPR is being performed and whether there is any deviation from optimal quality, and deviation occurs very frequently. This is the area that Dana Edelson focused on in her study. Edelson used a special defibrillator made by Phillips that collects and stores data during CPR. The machine provides instant feedback and also allows doctors to review the entire process afterward. During CPR, the machine gives voice commands, such as telling the person to press deeper or to allow more time between compressions. Other manufacturers, such as ZOLL, also have these technologies available, but most people don't routinely utilize them.

Edelson found that the rate in her hospital for restarting the heart was initially 45 percent, but after she retrained the doctors about the CPR basics and used the feedback system, the ability to restart the heart increased to nearly 60 percent. She also found that there were problems with the ventilation rate. In only 38 percent of the cases were doctors delivering the appropriate ventilation rate; in the remainder of the cases, they were actually either underbreathing for the patients or overdoing it, creating the phenomenon of breath stacking, a major factor in preventing the heart from restarting. Edelson managed to teach the staff how to better deliver breaths; even then, however, in half the cases the staff still did not deliver correct ventilations—not perfect, but much better.

Knowing that the quality of chest compressions and their delivery is so variable, many manufacturers have developed machines for this purpose. Several different machines aid in CPR, including Life-Stat, Lucas, and ZOLL AutoPulse. The Lucas works by a piston that compresses the chest. The ZOLL AutoPulse is a battery-operated machine that straps to the chest and delivers measured chest compressions. The LifeStat is a dual-functioning machine that handles chest compressions and breathing. Because its pumping action is caused by pressure from an oxygen tank, the machine takes the excess oxygen and acts as ventilator, so it ultimately delivers quality chest compressions and quality ventilation and avoids the problem of breath stacking and death by hyperventilation. It also takes just a short time to put it on the patient.

Another machine that is very helpful is called an extra-corporeal membrane oxygenation (ECMO) device. Basically, ECMO functions like an artificial heart and lung machine and delivers oxygen until we can get the heart restarted. This is used when the heart has stopped and therefore the blood cannot be oxygenated. To avoid further impending injury to the organs, the patient is hooked up to the ECMO, which takes the person's blood out of the body, passes it through the machine (where it is artificially supplied with oxygen), and then pumps it back into the body. This has become widely used in Southeast Asia, particularly in South Korea and Japan, where doctors have restarted the heart up to 70 to 90 percent of the time versus the usual best rate of 20 to 50 percent in the United States and elsewhere, but there is no across-the-board standardized procedure for its use in the United States or in Europe. Like much of resuscitation, there are variations in practice.

If the heart's electrical rhythm has short-circuited due to a heart attack or any other cause, it can be shocked back into rhythm. However, this must be done quickly. Each minute of delay in shocking the heart increases the risk of not being able to restart the heart by 5 to 10 percent. That's why defibrillators are so widespread in public places these days and everyone is encouraged to learn how to use them as well as how to perform chest compressions. Starting treatment early makes a big difference.

But restarting the heart is only the beginning. Critical steps must be taken to ensure the heart does not stop again. The blood pressure must be restored to provide proper oxygen to the brain. This is critical to resuscitating a patient. The outgoing pressure, called the systolic pressure (the high number when you have your blood pressure taken), is created by the heart pushing blood, and the return pressure, called the diastolic pressure (the low number), is created by the elasticity of the arteries. Even if the systolic pressure has come back, the diastolic must also come back as it is during this time that blood gets into the brain and heart.

As the cardiac arrest continues and a person's brain becomes more swollen, we need higher pressure to get more blood into the brain. So now it becomes an issue of not only reaching but exceeding normal blood pressure to pump enough blood into a swollen brain. If you can only get a minimal amount of blood into the brain, you are in a low-flow state. Studies using monitors called EEGs, which measure electricity coming off as a marker of brain function during cardiac arrest resuscitation, have demonstrated that the brain remains in a flatline state for some time even after the heart has been restarted and a person appears to have a normal blood pressure. This also answers the question of why so-called near-death experiences (or what I call actual-death experiences when they occur during cardiac arrest) are unlikely to be explained simply by chemical changes going on in the brain during cardiac arrest or even soon after, because the brain is not functioning. It can take a few hours for the first flicker of electrical activity and function to resume in those damaged brain cells.

In a 1996 study in cats, Dr. Malcolm Fisher of the University Hospital Bonn in Germany and his colleagues stopped the heart in fourteen animals for fifteen minutes and then performed CPR. Next, they added high doses of epinephrine and then shocked the heart back to normal. The result was that the cats' average pre-cardiac arrest blood pressure was normal at 107 and their cerebral perfusion pressure (the pressure of blood that gets to the brain) was 101. This is roughly what it would be in human beings as well. When they stopped the hearts of the animals, performed CPR, and administered a high dose of epinephrine, the blood pressure went up to 65–77, yet the cerebral perfusion pressure was only 37, or one-third of what it should be. Again, this shows us that despite everything we do, the patient is still in a very low flow state, and there is very little we can do to get enough blood into the brain.

THOUGH NEGOVSKY FIRST DESCRIBED postresuscitation disease in 1972, it wasn't until 2008 that it became much better understood and the first international guidelines regarding the optimal medical management of postresuscitation disease were published. I learned about them when I had coffee with the prominent resuscitation expert Dr. Jerry Nolan in December 2008. I was working in London at the time, and I met him to discuss research I was doing on the brain during cardiac arrest. He mentioned that new guidelines for resuscitation published jointly by the American Heart Association and the European Resuscitation Council, as well as the Australian and New Zealand Council on Resuscitation, the Heart and Stroke Foundation of Canada, and the Resuscitation Councils of Asia and South Africa, were coming out in a few weeks, and he was convinced that if they were put together properly, they could minimize brain and organ injury during cardiac arrest. The mere fact that I had learned about these over coffee underscores one of the gaping deficiencies of the system: although these guidelines are published, they are not systematically disseminated or followed because doctors don't automatically become notified about their existence and no one necessarily enforces them.

These 2008 guidelines, along with other comprehensive ones published in 2010, show both what doctors in the resuscitation community are doing well and what they could do better to improve survival rates. The areas of study can be divided into two periods. The first and most commonly known period—the time during which the heart has stopped and we are trying to restart it—is called the intracardiac arrest phase. For Joe Tiralosi, this lasted forty-seven minutes. For Arun Bhasin, this was more than three hours. During this period, if doctors perform CPR properly, they can often restart the heart. But after the heart is restarted comes an even more pivotal period, the postresuscitation phase.

One of the most significant understandings of the last ten years is that a primary reason cardiac arrest survival rates are so low is that not enough attention has been paid to the postresuscitation phase, the twenty-four- to seventy-two-hour period after the heart has been restarted. While doctors may celebrate after restarting someone's heart, the celebration is short-lived if the patient ends up dying the next day. In the seventy-two hours following the cardiac arrest, the probability of another cardiac arrest occurring is high, because there is massive injury going on. The focus on the postresuscitation phase is a seismic shift that is going on now. We now know that you can't just pay attention to one phase; you must focus on all phases together. This new approach is going to yield significant improvements in survival outcomes of those who suffer cardiac arrest.

The most important discovery (and a central part of the 2008 and 2010 guidelines) that helps in the postresuscitation phase is hypothermia, which buys time during a cardiac arrest. The recent increase in the use of cooling has emerged in response to doctors and researchers doing some soul searching. By identifying the problems, we can move to correct them and improve survival rates.

Studies show that the reason for death in the postresuscitation phase is evenly split among three causes: ongoing brain injury, ongoing heart injury, and massive inflammation in the body that causes chemicals to break down the organs. These three causes arise as a complication of the efforts to resuscitate people back to life after death has started. There is also a fourth vital factor that cannot be ignored: the original issue that caused the person to die in the first place. If this is not fixed, then of course it will continue to cause organ and cell damage and hence death again and again, until it is remedied or doctors give up on resuscitation. The main issue in the brain is swelling due to a prolonged period without adequate oxygen, called cerebral edema. As the swollen brain presses against the skull, the brain starts to squeeze the brain cells. This results in a secondary or delayed ischemia where adequate blood is not delivered to the brain for the next twenty-four hours. So even though the heart has been restarted and the patient is alive, if doctors examined the brain, they would find that a grossly inadequate supply of blood is being delivered to it. The result is that the brain cells (which take around eight hours to die) will continue going through the dying process.

Once we recognize the enemy, we can treat it. Because the brain is swollen, we have to elevate the blood pressure above normal to overcome the swelling and get blood into the brain. The other thing we have to do is make sure the patient doesn't have a seizure because we are trying to reduce the metabolic activity of the brain cells. If the cells are overactive and don't get oxygen to meet their requirements, they will die sooner, which is what happens when a person has a seizure after initially recovering from cardiac arrest. So what we do is hibernate or slow them down through cooling.

Cooling does a number of things. First, it reduces the swelling and pressure. It also reduces the activity of the cells, thereby shutting down the catalysts that drive the process of cell death. The initial impact is like putting the brakes on the process by which cells are dying. Therefore, in a situation where we are only able to deliver roughly 50 percent of the required oxygen (due to the physical resistance caused by a swollen brain), we must decrease the requirement of brain cells for oxygen, so that what we can deliver (even if it is not much) becomes equal to what the brain cells need for their metabolic requirements. This is the key factor that can save the brain cells—to match the blood and hence oxygen supply to their needs. This allows those cells that are semihibernating to get enough oxygen to stay alive and move off the path to death.

This treatment has now become recognized as a great savior for the brain after cardiac arrest. This was illustrated recently by the case of an emergency physician, Dr. Kelly Sawyer, who was researching the effects of cooling after cardiac arrest as part of her master's degree thesis at Virginia Commonwealth University. She also worked as a fellow in emergency cardiac care and was part of a team, established by Dr. Mary Ann Peberdy and Dr. Joseph Ornato, that provides hypothermia in the community and resuscitates patients and brings them to the hospital. In the final month of her fellowship, Dr. Sawyer underwent knee surgery, which is a risk factor for developing clots. She began experiencing pain in her leg but did not realize it was a clot. Then in June 2011, she was walking into the department on the day that was the deadline for submitting research abstracts to the national American Heart Association meeting. In the corridor between the parking lot and the building, she collapsed. Luckily, two nurses who were arriving for work called for help. The EMS service arrived quickly and drove her around the corner to the ER.

The two emergency room doctors who had been wondering why Kelly was late and were waiting for her to discuss her research were notified that she had suffered a cardiac arrest, and they rushed to the ER to help resuscitate her. They called the cardiology team, which performed a bedside ultrasound and found a blood clot sitting in the heart, which was blocking the circulation of blood and hence the flow of oxygen through the lungs and heart. This is typically potentially fatal, as stopping circulation due to any blockage in the system causes the heart to stop due to the mechanical obstruction it causes to blood flow.

By this stage, Sawyer was in and out of consciousness. Her heart stopped pumping blood a couple of times, and the emergency physicians performed chest compressions to revive her. They nearly put her on the ECMO machine (a type of heart-lung bypass device that is often used in Japan but rarely in the United States for cardiac arrest), but they managed to transiently stabilize her enough without it. They then sent her to the operating room, where they opened her chest and found the large blood clot, called a saddle embolism, that was blocking the flow of oxygen to the lungs and the blood to the left side of the heart. She was then cooled for twenty-four hours in the ICU to ensure there was no cell damage. She was warmed gradually. Six days later she was taken off sedation. Kelly became one of the few doctors whose work may actually have contributed to saving her own life and her brain. Initially, she noticed that she had been impacted and that she wasn't stable enough on her legs. After fourteen days, she was discharged from the hospital. Gradually, through physical therapy she was able to fully recover and I met her for the first time at the national conference where she was presenting her work on the effects of hypothermia to physicians and scientists from around the world. Incredibly, if it hadn't been for the effects of hypothermia, it is likely Kelly's brain would not have recovered.

In the postresuscitation phase, we must also deal with a weakened heart. During a cardiac arrest, the heart has a massive attack and its cells do not receive adequate oxygen for minutes to hours at a time. The result is that the heart muscle becomes stunned and does not squeeze properly, so even if you restart the heart, it is often very weak and not pumping the blood strongly enough. Consequently, we need to give specific drugs that can make the heart beat more strongly in order to raise the blood pressure. Once we recognize this, we can give the patient specific drugs to help the heart over the twenty-four hours it needs to recover.*

An interesting footnote to this is that in France in 1997, researchers conducting a postmortem study found that almost 80 percent of people who suffered a cardiac arrest had undiagnosed heart disease that had caused them to have a heart attack. These people

---

*When faced with critically low blood pressure and clinical shock, physicians and nurses usually prescribe blood-pressure-elevating medications that mainly work by squeezing the blood vessels and thus raising the blood pressure. These are most helpful in cases when a critically ill person is in shock due to conditions such as overwhelming infection and anaphylaxis, because in these cases the heart itself is not usually weak and the problem is that the blood vessels are too dilated and leaky due to massive inflammation. However, after a cardiac arrest, the problem is that the heart is weak and there is also massive inflammation; the cause of shock is a combination of the two. Oftentimes, the potent medications that people use to raise the blood pressure (norepinephrine, phenylephrine) are the same as the ones used in overwhelming infections and anaphylaxis. However, these drugs actually make the work of the weakened heart more difficult, and this is frequently unrecognized. Therefore, in these instances, the potent drugs given to increase the blood pressure by squeezing the blood vessels make the situation worse. The impact of squeezing the blood vessels alone paradoxically keeps on dropping the blood pressure as it makes the work of the weakened heart much harder and the heart soon fails as it can't push against the resistance of the tightly squeezed blood vessels. The blood pressure drops lower and eventually the heart stops again. The missing link in this instance is that without giving drugs that actually cause the heart to squeeze and beat more forcefully, the blood pressure won't improve. It is then often assumed that the person could not have been saved and the resulting critically low blood pressure and shock are normal consequences of the cardiac arrest that can't be improved. A decision may then be made to advise families that they should consider withdrawal of life support so that the person may die peacefully because care has now become futile. This is another illustration of the complexities of caring for patients with postresuscitation disease and the need for teams of experts in dealing with these complicated medical cases. However, again this is not required by legislation and thus is not usually implemented because it is simply unrecognized by hospitals that sincerely believe they have done all that they could.

had blockages in the blood vessels to their hearts that had not been recognized until their hearts stopped. Because of this, the standard of care in many countries now dictates that unless there is an obvious reason that someone's heart has stopped, the patient is taken to the catheterization lab for a cardiac catheterization. If this is not done and there is an underlying blockage, even if the heart is restarted, the person will most likely still die later. However, it isn't being done universally, and many institutions have their own criteria for which patients they will take to the cardiac catheterization laboratory. As a general rule this is done more readily in Western European countries and less in the United States, where doctors are more conservative in their approach. There is often a legitimate concern raised by the U.S. doctors that if patients are taken to the catheterization laboratory who are "already dead" and thus have less of a chance of being revived, then the governing bodies may actually penalize the doctors and even close down the catheterization laboratory at their institutions. This is because their overall annual mortality figures will potentially appear to be very high. The problem lies with the system, because when evaluating overall annual mortality data for each hospital, the state governing bodies in the United States do not always distinguish between unfortunate deaths that may take place in stable patients taken to the cardiac catheterization laboratory and patients taken after a cardiac arrest who had already died and were undergoing the procedure in order to potentially give them their lives back.

Inflammation in the body can also be a cause of death in the postresuscitation phase. Because the whole body is swollen, it is crying out for adequate levels of oxygen. What we have to do is measure the adequacy of oxygen delivery to the organs. If oxygen is underdelivered, the organs can go back into shock and die a second time. Therefore, the oxygen delivery must be restored in the first few hours so that organs don't become permanently injured.

Numerous studies have shown that cooling slows the deterioration of the cells in cardiac arrest patients and prevents the process of death from going too far down the line. The first study that addressed therapeutic hypothermia was published in 1959. It showed that patients who had survived a cardiac arrest (but remained in a coma) and who were cooled to 31°–32° Celsius (89°–89.5°F)—even when doctors waited between three hours and eight days to cool them because they were unsure how to proceed with resuscitation— had a good recovery 50 percent of the time, whereas those who weren't cooled had a good recovery just 15 percent of the time.

Throughout the 1960s, cooling was used in cardiac arrest cases, and many of the medical guidelines at that time directed doctors to cool patients with ice. But because they didn't know how to do it properly, sometimes the cooling did more harm than good; therefore, doctors stopped using it. The main problem was that using ice to cool the body carried the risk of overshooting the target temperature. When the body temperature dips below 32°C (89.5°F), there is a risk of complications. People can also experience more intensive bleeding or changes in the electricity of the heart that cannot be reversed. So like any treatment, if it is over- or underutilized, it may not be as effective.

In 2002, two comprehensive studies were published in back-to-back editions of the *New England Journal of Medicine* that both showed the benefits of cooling after the heart was restarted. The pivotal one was conducted by Dr. Fritz Sterz and his colleagues in Vienna, Austria. They looked at 275 people who had their hearts restarted after a cardiac arrest. The people were divided into two groups, 137 who were cooled and 138 who were not. What they showed was that 55 percent of the people who were cooled had a good neurological outcome, as opposed to 39 percent of those who were not cooled. They also found that the mortality rate after six months in those who were cooled was 41 percent versus 55 percent in those who weren't. Based on their research, for every six people cooled, one person benefited.

It's important to understand that by scientific standards a study with 275 patients is very small. Although the study showed that patients benefited from the cooling process, a larger study would probably demonstrate better outcomes.

Today, two main sophisticated cooling methods are used to remove the guesswork caused by ice. The first involves pads that are attached to the body with cold water circulating through them. These pads have an internal thermometer so that once the body temperature drops into the 32°–34°C (89.5° to 93°F) range, the pads maintain that range. The second method is inserting catheters inside the body. The catheters are placed in blood vessels so the person is actually cooled from the inside, which is even better. This method cools more quickly and regulates the temperature more accurately. Another novel yet less commonly used method involves cooling people by spraying a rapidly evaporating coolant liquid directly into the nose. This area of the body acts as a heat exchanger and cools the brain as it lies right under the brain. But as proved by the first few hours of Joe Tiralosi's case, ice bags can still work for a few hours if the modern methods aren't available.

If cooling is incorporated into a comprehensive treatment program with other postresuscitation phase treatments, survival rates will greatly increase. This was shown in a study carried out in 2007 by a group of researchers in Norway. In the study, the researchers didn't pay attention to the intracardiac arrest phase, meaning the quality of chest compressions, ventilation, or drugs administered; rather, they just focused on the postresuscitation phase. They decided that irrespective of the quality of the techniques that their medical colleagues had used to restart the heart in the emergency room, they would now focus on minimizing the effects of the postresuscitation disease. They wanted to improve the terrible statistic in which two-thirds of the hearts that had been restarted would stop again. They first instituted a system to treat patients in the postresuscitation phase. When a patient came to ICU, the researchers used a protocol of steps to be taken. In short, they formalized the approach to postresuscitation care by enacting all the different treatments that had been shown to work. They started with hypothermia. Every cardiac arrest patient was cooled to slow the process of cell death. They took all the patients who didn't have any other obvious cause for their cardiac arrests to the cardiac catheterization lab to check for unknown blockages in the blood vessels of the heart, and they controlled the oxygen and carbon dioxide levels, as well as the blood pressure.

The results were enlightening. Over a two-year period in a study of 69 patients, the survival rate increased from 26 percent to 56 percent. Further, 91 percent, or 31 of the 34 survivors, had a full and complete neurological recovery, whereas before it was 50 percent. So not only did the overall survival rate more than double, an overwhelming number of the survivors did not suffer brain damage because of the procedures implemented in the postresuscitation phase.

COOLING IS THE MOST revolutionary advancement in resuscitation medicine over the last twenty years. Scientists have determined that organs and cells die at varying rates depending upon their resistance to a lack of oxygen. The kidney is more tolerant than the liver, and the human cornea, for instance, can yield viable cells up to seven days postmortem. Because this deterioration of the cells due to a lack of oxygen is a dynamic process, it can be slowed down and even reversed—thereby allowing death to be reversed.

Many deaths can be reversed through cooling. Take a child who falls down on the basketball court and dies. If there were cooling devices in the emergency response vehicle, the paramedics could cool the child immediately and rush him to the hospital where doctors could determine what stopped the heart and correct the problem— thus saving the child's life.

The entire process becomes a function of how much time is needed to fix the underlying cause. If the problem can be corrected in a matter of minutes or in some cases hours, we can now rescue that person from the clutches of death. Future advancements in medicine and technology promise even more. In fact, within the next ten to twenty years, we might be able to recover loved ones many hours or even days after they seemingly breathed their last breath, repair the underlying cause, and ensure they go back to a normal life.

The bottom line is that if the cooling process and postresuscitation care is done correctly, the patient can come back to life without brain damage. Cooling and optimal postresuscitation care is one of the dividing points between those who suffer brain damage after cardiac arrest and those who don't. If cooled, all those cells that were deprived of oxygen can again return to normal.

In June 2001, the Department of Emergency Medicine, Gunma University Graduate School of Medicine Maebashi in Gunma, Japan, reported on a case that illustrates the successful use of cooling combined with the ECMO machine. The patient was a thirty-year-old female who was found in a forest at 8:32 A.M. following overnight exposure to cold due to an overdose of medications. She was dead. Her body temperature had dropped from 37°C (98.5°F) to 20°C (68°F), meaning that she had been there for several hours, as the body temperature drops by about one to two degrees per hour after death. The medical team arrived at 8:49 A.M., administered CPR, and shocked her heart using an automated external defibrillator, but she remained dead.

When the woman arrived at the hospital at 9:22 A.M., her body temperature was still 20°C (68°F), and her pupils were equal in size and not reactive to light. The ER doctors performed CPR and inserted a breathing tube and ventilated her lungs with an automatic ventilator, all while continuing chest compressions. The drugs adrenaline, amiodarone, and lidocaine were injected to restart the

heart. Despite efforts to begin warming her up, the woman's temperature remained unchanged. The doctors then hooked her up to
the ECMO machine to ensure optimal oxygen supply.* After six
hours of treatment, her temperature returned to 32°C (89.5°F), and
her heart restarted. Although she had remained physically dead for
a further six hours after being discovered dead, having died sometime the night before, the patient was able to recover and eventually
walk out of the hospital without organ and brain damage. Because
she had been naturally cooled down by the environment during the
time that her heart was stopped, her cells did not sustain the same
degree of permanent damage as those of someone who dies in a
warmer environment and were able to return to functionality once
oxygen delivery was restored.

While there is a lot that can be done today to reverse death, the
future is perhaps even more fascinating. In the American Heart Association conferences I attended in 2010 and 2011, there were pre-
sentations about developing special infusions of oxygen molecules
that are placed into the middle of a larger molecule that can carry
oxygen to all the organs even without the lungs and heart working.
Basically, researchers are putting oxygen in the middle of a super
tiny fat globule and injecting it. By using millions of these exceed-
ingly small fat globules that have oxygen molecules in the middle of
them, doctors may be able to artificially oxygenate people by injecting them with this special solution after they have died. We can't
directly inject air into the blood, but if we inject the fat solution
into the bloodstream, when it gets into the lungs and other organs
such as the brain and heart where the capillaries are very small, then
the fat globules open up and release the oxygen and thus supply
the organs with oxygen and slow down the rate of cell death after
a person has died. This is being tested in animals, and the studies

---

*This technology is used more frequently in Southeast Asia than in the United States
or many European countries.

are promising that it may be done on humans. This is part of what the near future may hold. If doctors are able to cool patients and inject them with a solution that supplies oxygen artificially, especially if given with another injectable drug that can actually block the action of the chemical enzyme pathways in brain cells that lead to death, then brain cells will not rapidly careen toward their own deaths even if the patients have died. In essence, the individuals will remain in the gray zone after death and avoid irreversible permanent brain cell death for a much longer period of time.

Two studies conducted by Dr. Robert Neumar, currently chair of the University of Michigan Medical School's Department of Emergency Medicine, and his colleagues offer a glimpse into the future of how we might be able to slow down the rate by which the chemical catalysts or enzymes in the brain steer brain cells toward their permanent deaths. In both studies, the researchers concentrated on the enzyme calpain, which is involved in the process of apoptosis that leads to cell death after a person dies.

In the first study, Neumar and his team hypothesized that if they were to block the calpain enzyme pathway from acting, they would be able to slow the amount of irreversible brain damage that takes place in rats that have been exposed to a lack of oxygen. In one group of rats, they cut off the oxygen to the brain for ten minutes, essentially simulating what would happen in a cardiac arrest. Next, they infused the rats with a drug that blocked the activity of the calpain enzymes. In a second control group of rats, the researchers gave a compound that did not have the drug in it. In the rats given this direct blocker of the enzyme, it was found that seventy-two hours later brain damage was markedly reduced. Therefore, if researchers could develop this for human beings, we could potentially infuse this as a drug, and it would then reduce the amount of brain damage during and after a cardiac arrest and slow down the rate of progression from reversible to irreversible brain cell death after someone dies.

In the second study, also conducted in rats, Neumar and his colleagues gave a different form of the molecule that inhibits the activity of the same group of enzymes, the calpains, in rats who had been left for a period without oxygen being supplied to their brains. What they found was that those rats with the drug had increased survival of their brain cells. They took this a step further by getting the rats to perform certain activities. They found that the rats were able to function better because they had had preservation of certain areas of their brains.

Both studies isolated one enzyme and showed that if we can target that enzyme, then we may be able to reduce the rate at which brain cells go from reversible to irreversible death. This is therefore a possible avenue for drug treatments that can preserve the brain after people have died. But in the practical world, and focusing on today rather than tomorrow, the biggest problem is the dissemination of high standards of care based on what we already know, as detailed in this chapter. In fact, more comprehensive implementation of optimized medical care aimed at improving survival and brain injury from cardiac arrest has proved to be very difficult because there is little coordination between the disparate areas of medicine involved. The discoveries outlined here will not lead to better overall survival rates and lower the occurrence of brain injury unless there is close coordination between the medical disciplines involved and unless the good quality of care needed is actually implemented. Otherwise we will forever live with a zip-code lottery type of care, where we have to hope and wish for the best when the inevitable happens to us.

CHAPTER 5

The Orphan

CHARLES LINDBERGH'S HISTORIC FLIGHT across the Atlantic Ocean was triggered in part by a bet. New York hotelier Raymond Orteig offered $25,000 to the first person who could fly between New York and Paris, and several pilots lined up to take the risk. While Lindbergh's successful transatlantic flight in the *Spirit of St. Louis* in May 1927 has been well chronicled, it is little known that four of his competitors died. Two were killed in a test flight in Virginia, and two others attempting the journey from France disappeared after crossing over Greenland. Clearly, such a journey in a plane made of wood strips held together with wire and linen was fraught with peril.

Domestic commercial air travel at that time was nearly as dangerous. Though the planes were sturdier, there was still not a comprehensive system in place to enable flights to take place safely. For example, there wasn't the same level of integration between pilots and crews on the ground, airports were primitive (often not much more than a large field), and communication was limited, as was the understanding of how to handle safety issues on board. For instance, if the seal on the door were to become compromised, the pilot would likely continue the journey rather than going through a checklist and then making adjustments in altitude, cabin pressure, and flying speed to prepare for an emergency landing. Consequently, passengers boarding a plane were almost literally taking their lives into their own hands. In the United States from 1926 to 1927, there were twenty-four fatal commercial airline crashes. Things got worse in the following two years as commercial airlines logged sixty-seven deadly crashes. The fatality rate was one passenger death for every three thousand flying—which would translate to seven thousand deaths per year in today's numbers. This is staggering given that from 2000 to 2010, even with the deaths on 9/11, the fatality rate was just one in eight million.

The aviation industry in its infancy was roughly where cardiac arrest care and resuscitation are today in terms of the implementation of optimized system management, as well as, of course, safety measures. Despite the fact that planes were becoming more reliable throughout the 1920s, there was a lack of overall system management in place, so fatalities remained high. Likewise, nascent resuscitation techniques discovered in the mid-1960s, such as closed chest compressions and artificial breathing through mouth to mouth, were being done in isolation and not as part of a comprehensive and integrated system of care with attention being paid to every vital link of what we have termed the "chain of survival." But once aviation married better equipment with a comprehensive and systematic program that included better aircraft design, materials, aerodynamics, landing gear, on-the-ground inspection teams, specialized airports (instead of fields), radar, crew management and teamwork, and a checklist safety system, fatality rates dropped rapidly—to 1 in 450,000 in the 1960s and then to 1 in 2 million by the 1980s.

Aviation and resuscitation science share many similarities. Both are new discoveries that have enabled human beings to overcome something that had forever seemed impossible—a flight of fantasy or a dream that would never be within the realm of possibility. What aviation and resuscitation science share most of all is that they both require a highly complex, sophisticated, and integrated system of management, which if it fails inevitably leads to fatalities or devastating disabilities—even if the failure was due to just one error.*

Today we take flight for granted. It doesn't seem so novel anymore. When we sit on a modern aircraft and fly tens of thousands of feet above the ground and thousands of miles across the earth, we are most likely not even aware of the human toil, sacrifice, perseverance, and dedication that eventually led to this truly incredible achievement. But in 1903, after many attempts, the human dream of flying using a machine heavier than the air was finally realized on a remote, windswept beach in North Carolina. In truth, even that event itself was far from earth shattering. A "flying machine" built by two brothers managed to overcome gravity and fly a little above ground for only twelve seconds—a far cry from what we have achieved today but nonetheless incredibly significant. From this small event, now we overcome the force of gravity using a machine that is far heavier than air and fly anywhere we want, including into deep space. Anyone who has seen the sketches of Leonardo da Vinci's flying machine will get a feel of what a real fantasy this idea must have been to our forefathers. But in just over sixty years after the Wright brothers' first flight using their primitive plane, we developed aircraft and engines that could not only safely fly thousands of feet above the ground, but could also fly supersonically and even hypersonically reaching the highest points of the stratosphere and even on to the moon. The challenges were incredible, but what was achieved in this time period simply attests to the incredible human resolve in overcoming any challenge, no matter how seemingly impossible.

This comparison has been put forward by many, including Drs. Joseph Ornato and Mary Ann Peberdy from Richmond, Virginia.

By contrast, in the almost sixty years that followed the discovery of modern resuscitation by pioneers such as Safar, Jude, Knickerbocker, and Kouwenhoven, we have not witnessed anywhere near the same level of progress in this field. In fact, despite the fact that we have made enormous progress in the fields of science and medicine, and we now have cardiac catheterization laboratories, modern ventilators, and powerful blood pressure drugs to keep people alive, our long-term survival rates from cardiac arrest have not particularly improved in this time frame. It's really amazing but absolutely true. When I first came across this statistic, I frankly didn't believe it. I thought it was impossible. I actually searched the original studies from the 1950s and 1960s up to the present day but found that it was indeed sadly true.* Interestingly, when I now bring this up at conferences or lectures, it has the same jaw-dropping effect on others too. In many ways, resuscitation science reflects the opposite to aviation. That is, although resuscitation after cardiac arrest is an incredibly innovative scientific discovery and the field has progressed scientifically in the past sixty years, it has lacked a sophisticated and coordinated system of management. Resuscitation science reflects the story of at best limited and often failed universal implementation and adoption of the highest standards combined with a lack of external regulation. It also reflects what happens to a medical condition that cuts across all boundaries and is thus not "owned" by any one group of medical specialists such as cardiologists, neurologists, or emergency medicine doctors. Consequently, no one group drives the highest standards of care in our hospitals and in the communities.

Asthma has a home in pulmonary medicine. Cancer is the domain of oncology. Parkinson's belongs to the neurologist. But cardiac arrest is an orphan by virtue of the fact that it cuts across

---

*Clearly, pockets of improving care do exist, but in general, outcomes have not improved.

many specialties because it is death, and death happens in all specialties of medicine but is parented by no one. This comes as a surprise to most nonmedical professionals, who assume cardiac arrest and a heart attack are one and the same, but as we have demonstrated, they are not. Cardiologists specialize in the heart attack, which occurs when a blocked blood vessel stops blood flow to the heart, but they are generally not trained in the full spectrum of resuscitation because this requires expertise in a whole lot more than just the heart. It includes many delicate noncardiac facets involved in restarting the heart combined with excellent postresuscitation treatments, which should be the area of responsibility of the intensive care physician, as well as brain management, which should be the province of specialized neurological intensive care physicians. A cardiologist is not an expert in critical care medicine, including the management of complex lung and ventilation disorders, while the intensive care physician is not trained in the nuances of cardiac care, and neither of the two is usually fully versed in the intricacies of brain management. Of course, the neurological intensive care physician who is an expert in this area is also not an expert with the heart and lungs.

This in itself leads to variations in treatment. For this reason, even in the same institution, the same patient will not infrequently receive different care depending on who is in charge and which ward the person is taken to, such as the medical intensive care unit, coronary care unit, surgical intensive care unit, or neurological intensive care unit. This applies to the large institutions with multiple intensive care units, while in smaller community hospitals with fewer resources the care can also vary. Furthermore, inadvertently people may manage cardiac arrest without clearly following all the latest recommendations, discoveries, or guidelines because they may not be fully aware of their existence, or even how to implement them, which also contributes to the enormous variations in the standard of care. It is seeing this variation that drives me to push on and work hard in this field, and it is this that motivates me to write this book and highlight some of the areas that need to be improved in order for us to save many more lives and brains.

Today, when it comes to safety, there has been a reversal in aviation, and most of us feel completely safe on an airplane. Modern-day commercial air travel is statistically the safest mode of mass transportation—even flying across the Atlantic Ocean by way of Greenland. Airline crashes garner major national news coverage because of the sheer number of people involved, whereas even the worst automobile crashes are typically limited to a segment on the local news. But the fact is that flying is overwhelmingly safer than traveling by car. Consider that in 2008 the National Transportation Safety Board in the United States reported 1.27 deaths for every 100 million miles driven in an automobile. By contrast, there were no commercial airline deaths that year in twenty accidents reported; therefore, there were zero airplane deaths per 100 million miles flown. According to the National Safety Council, the odds of dying in a motor vehicle accident over an average lifetime are 1 in 85, or 1 in 6,584 per year, whereas the odds of dying in an airline crash are 1 in 5,862 over a lifetime, or 1 in 455,516 per year.

Flying is far safer than it was in Lindbergh's day because numerous advancements have taken place in aircraft design, navigational instruments, and safety procedures on the ground and in the air. But the main reason that flying is the safest mode of transportation is the development of step-by-step protocols for servicing and operating airplanes that include many people working together seamlessly in concert in all stages of a flight. Through these detailed procedures, mistakes can be found and corrected before they endanger lives both on the ground and in flight. The many perils of everyday air travel, such as foreign object debris, lightning, ice, engine failure, bird strikes, volcanic ash, misleading information, criminal acts, and human error, can be mitigated through a system of checklists and procedures that have been implemented by the airlines and are enforced by bodies such as the Federal Aviation Administration in the United States, the Civil Aviation Authority in the United Kingdom, and other similar organizations across the world.

These procedures involve a "chain" with many links, and if one or two links are missing, the results can be disastrous. Consider the deadly Air France Concorde crash in 2000 that ultimately resulted in the grounding of the Concorde altogether. Investigators found that several steps in the safety chain had broken down. The plane was loaded up as much as one ton over its maximum weight, and its center of gravity was tilted toward the rear rather than being centered. Further, fuel distribution was incorrect and fuel shifted during taxi and overfilled the number five fuel tank. Minutes before the Concorde took off, another plane in its path lost a small strip of alloy on the runway, but this debris remained because a mandatory tarmac inspection was not completed. As the Concorde was taking off, it ran over the debris, which was then slingshot into the number five fuel tank, causing a shock wave that resulted in the fuel tank rupturing. The fuel that spewed out was ignited and the fire damaged one of the wings, rendering the aircraft unstable in flight and ultimately causing it to crash.

To better describe the state of resuscitation, imagine if today the risk of flying had always remained as high as in the 1920s and everyone simply took this fact for granted because "well, it has always been risky to fly." Furthermore, the standards of pilots and the various crews needed to provide the requirements for flight varied greatly not only from airline to airline but even within the same airline. However, no external regulation and monitoring ensured standards. Also in this operational climate, when new scientific discoveries took place that could help improve flight and make it safer, they would either just not be adopted or adopted piecemeal during some flights but not others. For instance, imagine if following the discovery of radar, some airports or airlines continued to work without it, or it was only used on some flights while on others it wasn't, because some people "believed" in radar while others didn't even though its benefits had been acknowledged scientifically and its use recommended by the various national and international aviation bodies.

Imagine further if external organizations did exist that could make recommendations, but there was no organization to enforce the highest standards whether it be the incorporation of radar, or the level of training for the staff involved in flight. So ultimately the decision to train staff or use radar or any other factor was left to individuals such as pilots, airline managers, or local airports, and their decision to adopt these or not would of course reflect a variety of issues such as knowledge, experience, expertise, financial constraints, and so on.

Although the preceding is thankfully largely unthinkable in the airline industry, this is the situation we face with resuscitation science. Without doubt, the majority of our doctors are remarkable and exemplary, as are our pilots. The expectation that as well as knowing how to fly, a pilot or any one individual should also know all there is to know about radar, aircraft design, materials, running an airport, creating landing systems, and so on, and have the final say about all these vital components involved in successful aviation would seem absurd. Medicine too has become so sophisticated that it would be impossible to ask a physician or any hospital manager to know all there is to know. This is why, even though I may meet a remarkable physician, it would be impossible to expect him or her to know how to manage all the nuances of resuscitation science, because in the same way that flight is much more than taking off from the ground for a short time using a simple aircraft made of wood, resuscitation is also far more than chest compressions, breathing, shocking the heart, and giving drugs.

So while we figured out long ago that success with aviation would only come about through the establishment of a universal system that puts together all the important components into preflight, in-flight, and postflight stages and thus made aviation the safest form of mass transportation around the world, we have not implemented the necessary complex systems of care needed to overcome the issues of precardiac arrest, intracardiac arrest, and post-cardiac arrest complexity and ensure acceptable outcomes with this condition. This applies to our hospitals, our ambulance services, and our health-care establishments. This is the big challenge we face today. It's like saying we know how to make better aircraft and we have great pilots, but we still have the same overall fatality rate associated with flight—the problem is not with the pilots or planes, it's just that we haven't incorporated all the other vital components, such as good airport landing systems, radar, crew management, and so on, into an efficient system.

Resuscitation science could learn from the airline industry. Pilots spend countless hours in simulators before they fly each aircraft. They practice crisis management in concert with flight attendants and emergency personnel, so when a disaster occurs, they have a detailed procedure to follow that eliminates much of the guesswork. In contrast, in resuscitation science, only a relatively basic training program, called Advanced Cardiac Life Support (ACLS), is in place for doctors. As mentioned earlier, it teaches basic teamwork and co-operation skills, as well as how to perform chest compressions, assist breathing, provide shocks, and give drugs correctly.

These courses have many limitations. For starters, the overall level of the course corresponds more or less with the core discoveries made in the 1960s and 1970s, and it doesn't put together all the other various complex treatments that are involved and needed for greater success with this condition. It teaches medical, nursing, and ambulance personnel how to deliver the basic core components that, while undoubtedly very important, reflect only a fraction of the level of knowledge needed. Although great as an introductory course to resuscitation science that teaches the basic skills required to restart the heart, which is vital, the course is too basic.

The other major problem is that these courses are not even always mandatory for medical professionals who could end up dealing with a cardiac arrest case. Some assume that by virtue of having trained in cardiology, intensive care, or emergency medicine a physician would simply *know* and should be able to deliver all the nuances of resuscitation. Patients who have suffered a cardiac arrest are by definition the sickest patients in any hospital, and many have multiorgan failure resulting from a lack of oxygen being delivered to their organs and are in a state of medical shock. They often have brain failure, heart failure, kidney failure, liver failure, lung failure, infections, and much more.

Training in any one field of medicine does not provide expertise with all the nuances of cardiac arrest cases, since the nuances cross many different specialties; the knowledge required is real hands-on practical knowledge and not theoretical knowledge. While we can't expect such a high level of knowledge from our ambulance crews, nurses, or junior doctors who may be the first to the scene of a cardiac arrest, they all need to be able to perform the basics very effectively and deliver patients to specialized units with a comprehensive system that can deal with all nuances. Currently these systems and hospitals do not exist in a formal systematic fashion. It is very easy to save a life in theory but very hard in practice. The courses are also designed to cater to all levels and thus often contain people of different abilities. For example, a nurse working in an outpatient skin clinic who has never dealt with a critically ill patient will do the same ACLS course as a senior cardiologist, a medical director of an emergency department, or a senior intensive care physician who deals with life and death every day. This simply highlights that the level of knowledge taught in these courses, while vitally important, is too basic and only touches part of the problem.

The ACLS course contains some simple role-play and simulation cases, but they are not extensive or detailed enough. In addition, for those who take the course, it is usually repeated every two or three years, depending on which country the person lives in, but realistically speaking it is not possible to expect people to remember all the details of any course, let alone an important lifesaving course with two- or three-year blocks of time in between attendance. Many studies have confirmed that significant skills decay within just a few weeks to months after taking the course. We had firsthand experience of this at my own hospital after we put together a training course for our emergency doctors last year and found that when we retested them a month later, they had forgotten much of it. As a result, we have now put together a monthly detailed resuscitation course called Resuscitation Plus. This is why pilots train and train for long periods in simulators before gradually working their way into actual passenger flights. Unfortunately, we don't have such an organized system in medicine. Some individuals assume that just having a medical degree or a particular qualification is sufficient, which it is not. It is just the beginning.

An analysis of a major U.S. academic medical center in 2011 showed that of all patients who had had a cardiac arrest and who should have received hypothermia treatment (based on the institution's own criteria), 40 percent had not been given this treatment, and the overall survival figures were lower for this group. When asked why they hadn't provided the care, many physicians simply either didn't know about the recommendation or were not aware of the weight of evidence behind it and thus stated they weren't sure if they believed in it, or simply felt they didn't have the experience or comfort level to deal with it. These physicians included cardiologists, emergency physicians, intensive care physicians, general surgeons, and cardiothoracic surgeons—a wide array of doctors—many of them highly accomplished and all at the same institution. Although the institution had an approved "hypothermia protocol" with clear recommendations regarding who should receive this treatment (as with many other places), it did not enforce the protocol's implementation. This returns to one of the main challenges: neither this institution nor any other is mandated to provide this care because no governmental regulatory body exists to enforce it. Ultimately it is left to individual physicians with different areas of expertise, different levels of knowledge, and different comfort levels. Managers also can't be expected to enforce a protocol, because they don't have the knowledge or expertise and they are not being asked to do so by regulatory bodies. In this case, receiving hypothermia seemed to relate to the location where a patient with cardiac arrest was cared for in the hospital, indicating a lack of universal understanding; the American Heart Association (AHA) and other world bodies give hypothermia their highest recommendation. Consider that this randomness does not occur in heart attack patients. The standard of care in heart attacks is to have the blocked blood vessel supplying the heart opened as soon as possible (ideally within sixty minutes to prevent long-term damage). Therefore, hospitals have door-to-needle and now door-to-balloon times (i.e., opening the clot by powerful clot-busting medication or through a cardiac catheterization with a balloon). This is because a heart attack has a medical parent (cardiology) and because it has been pushed onto the agenda by regulatory bodies that collect statistics regarding door-to-balloon or door-to-needle times and severely penalize hospitals if they fail to deliver. Imagine the uproar if 40 percent of heart attack patients in our hospitals were not taken to the catheterization laboratory or given the powerful clot-busting medication simply because some physicians didn't know about the recommendation or were not aware of the weight of evidence behind it and thus weren't sure if they believed in it or simply because they felt they didn't have the experience or comfort level to deal with it!

Because cardiac arrest occurs across specialties, combined with the long-standing perception that it is essentially the end and that not much can be done since outcomes have always been poor (like saying flight has always been risky), together with the fact that it does not belong to any particular medical group (such as cardiology), it means that patients sometimes do not receive all that may be available when considered within the full requirements set out by international guidelines. We have a real opportunity to learn from the airline system, because although cardiac arrest is complex, a workable system can be implemented if we break down all the stages and create a comprehensive system that incorporates all the details and nuances relating to precardiac arrest, intracardiac arrest, and postcardiac arrest care. With adequate resources and continued educational programs, as well as a detailed checklist system like that which the airlines use for preflight, in-flight, and landing, a comprehensive system could integrate these disciplines and have an external agency mandate things be done in a certain way in much the same way the governmental agencies oversee the airline industry. Clearly, there is a need for this, and although some action has been taken by certain groups within the medical community, little will be achieved without proper and dedicated resources as well as external regulation and enforcement that ensure implementation of the highest levels of care for all.

MEDICAL SHOWS HAVE BECOME very popular on television, and in shows such as *E.R.* and *Grey's Anatomy,* most of the time that doctors perform CPR they are successful. In one study, called "Cardiopulmonary Resuscitation on Television—Miracles and Misinformation," by Susan Diem at the Durham Veterans Affairs Medical Center in Durham, North Carolina, and her colleagues, published in the *New England Journal of Medicine,* it was found that the success rate of CPR on TV is far higher than in real life. While this might be expected for dramatic purposes, the fact is that TV shows have to bend reality because the real-life success rate of CPR is less than 20 percent in hospitals and a lot less out of hospitals—on TV the success rate was found to be almost 70 percent. What's even more staggering is that as mentioned already, this success rate has actually not really improved since the 1950s—an outrageous statistic given all the advances in medicine over the past six decades. So how could this be?

What exactly are those things that enable people to survive that need to be implemented? The American Heart Association has a "chain of survival" that can be analogized to a bridge. In the same way that a bridge must have every section in place for a person to get across safely, the same is true of resuscitation. Imagine trying to cross the Golden Gate Bridge in San Francisco with just one or two segments missing! The consequences would be devastating. In resuscitation, we often miss many sections, yet each section must be followed or the outcome will be compromised, and the chain works as successive studies described in international guidelines published by the American Heart Association, European Resuscitation Council, and other world bodies have detailed. In 2005, the overall rate of survival from out-of-hospital cardiac arrest in Arizona was 4 percent (11 percent for witnessed ventricular fibrillation—this is the type that responds well to shock treatment using a defibrillator and is usually the easiest to treat). Survival increased with the continued implementation of each link in the chain of survival, and in 2009, overall survival reached 10 percent (30 percent for witnessed ventricular fibrillation).

The chain of survival begins when help arrives. This is a factor no matter where the patient has the cardiac arrest, but in a hospital, doctors should be able to arrive within five minutes. Out of the hospital, it depends on the ambulance service, but again we should aim to arrive within a few minutes from a person making a call. Even though doctors should be able to arrive quickly in a hospital (or paramedics in the community), the results are still only going to be as good as the care. This starts with the quality of the chest compressions. Perfect chest compressions can only deliver between 25 and 30 percent of the circulation that takes place before the heart stops, hence at best, oxygen delivery may be only one-third of what it is in a person whose heart is beating. From the perspective of the cells in the body, this is not enough to prevent progression from a state of potentially reversible to irreversible cell death in the organs, but it is better than nothing and can slow the speed by which cells die down.

CPR is a terrible grind. Administering steady compressions in two-minute shifts is difficult even for those who are trained and in the best of shape. In practice, during the frantic minutes of a resuscitation effort, medical personnel take turns conducting chest compressions on the patient. Some people are more effective and better trained than others. The primary issue is that people giving CPR become fatigued, resulting in the consistency being thrown off. No one can generally deliver effective chest compressions for more than a minute or two. Numerous studies have shown that doctors, nurses, and ambulance crews alike cannot deliver the optimal quality of chest compressions even if trained.

In one study, it was found that when people highly trained in the delivery of chest compressions—in fact, instructors who teach the courses—were put in an ambulance, they were able to deliver effective chest compressions less than 40 percent of the time. In this study only the use of an automated machine enabled the delivery of greater than 90 percent effective chest compressions. So clearly if someone is doing inferior chest compressions, then the patient is much more likely to either just not survive or suffer brain damage because not enough oxygen is being delivered. If the quality isn't perfect, then by definition oxygen delivery will be less. While chest compressions should be continuous with no more than ten-second pauses every few minutes, studies have shown that very often the pauses are very long, or the depth or rate of chest compressions is not adequate. This is assuming the person delivering chest compressions is physically strong enough to do so. If somebody is five feet tall and weighs one hundred pounds and is trying to deliver chest compressions, it is almost impossible to sustain the proper level even for a minute. People believe that they are doing proper chest compressions and they are going through all the right motions, but they are not delivering them effectively; and since there is no standard-ized system to check the quality, doctors can be executing chest compressions without knowing how effective they are. This is one major segment that is almost universally missing.

Although we now have the technology to give us feedback re-garding the quality of CPR—which we did not have in the 1960s, 1970s, or 1980s—there is no requirement for hospitals or ambulance crews to use it. This is like a pilot flying without an altimeter. If the pilot is flying in the dark and doesn't know how high his plane is, then he might fly into a mountain. No pilot would take such a risk. So now that feedback systems for CPR quality have been devel-oped, they should be standard issue for everyone performing CPR, whether in hospitals or ambulances. Currently, they are only used to a limited extent by very few individuals or hospitals and often as part of research studies. But as Dr. Dana Edelson's study at the Uni-versity of Chicago showed, outcomes can be improved if feedback machines are used.

In view of these findings, we managed to convince administra-tors at my hospital to purchase a number of automated chest com-pression devices so that we could deliver the optimal quality of chest compressions combined with systems to provide feedback regarding our quality. Although there are three main automated chest com-pression devices, called the Lucas, the LifeStat, and the AutoPulse, on the market, few people actually use them. We brought in the LifeStat because in addition to chest compressions, the LifeStat also regulates breathing. Part of the problem is that not enough research has been done to convince hospitals to use CPR feedback machines or automated chest compression devices. The AHA guidelines pub-lished (but not enforced) for resuscitating a patient are neither for nor against the machines. This has left hospitals in limbo. We found that we were able to restart the heart 75 percent of the time using our machine compared to 45 percent when chest compressions were done by hand, and this improvement seemed to be directly related to the ability of the machine to provide higher oxygen delivery and blood flow to the brain, heart, and other organs. We are now trying to get the system in place everywhere in the hospital.

In CPR, the rate at which breaths are supplied is just as important as the chest compressions—too many breaths actually themselves can kill people due to the phenomenon called breath stacking that I referred to earlier. This is why some researchers have called it death by hyperventilation. In many cases, people frequently provide too many breaths, a fact illustrated in Dr. Dana Edelson's study in which she found that though the recommendation is eight to ten breaths per minute, in reality patients were often receiving thirty-five to forty breaths per minute. Therefore, without a system to regulate the number of breaths being delivered, some lives may not be saved and people will simply assume it was because the person had suffered a cardiac arrest. In resuscitation, attention to detail really counts. However, excessive ventilation is another almost universally missing section on the bridge to recovery. The LifeStat machine has the advantage of being able to deliver the exact quantity and quality of chest compressions and breaths needed and thus take the delivery of breaths out of the hands of humans.

During resuscitation, the other factor in the patient that may need to be addressed is an abnormality in the heart. This can only be treated if the heart is shocked. For every minute of delay in identifying those rhythms, mortality increases by about 5 to 10 percent. If this issue is not identified and treated, those irregular rhythms will eventually become a flatline, and a flatline is clearly much harder to treat than abnormal rhythms. Modern defibrillator machines are available, such as the R series made by ZOLL, which allow doctors to see the exact heart rhythm even when chest compressions are ongoing (called see-through technology); they thus provide shock treatment immediately and without delay when it is needed. Again, these systems have generally not been adopted so far, and so in reality there is usually at least a few minutes of delay even at the best of

times. At our hospital we are trying to incorporate these technologies in order to provide a seamless and completely automated system
of care. We have also tried to reach out to our community ambulance services to implement some of the lessons we have learned in
the hospital, so that we can all work together to create a streamlined
service and system that focuses on quality from the moment the
heart stops until a patient arrives at the hospital and then through to
discharge from the intensive care unit.

In the United States, ambulances can be a bigger issue than
many people realize. Suffolk County, New York, where I work,
is one of the most affluent counties in the United States, yet it has
one of the worst out-of-hospital cardiac arrest survival rates at 2 to
3 percent. Part of the problem is that literally a hundred different
ambulance companies are providing service, much of which is provided by highly dedicated yet volunteer crews who get paid from
other jobs. Learning about this was an eye-opener for me because
I originally grew up in London and then lived in New York City
before moving to Suffolk County. Neither city runs a volunteer
service. In fact, I had never heard of or even considered that an
emergency service dealing with life-threatening situations such as
cardiac arrest, where every minute counts, could be manned by volunteers, but this is what happens in our region. In fact, the county
has recruitment billboards for fire and ambulance emergency personnel that boldly states "work for pride not a paycheck."

The question is this: How do we regulate personnel who, although amazing and remarkable individuals by virtue of the fact
that they have selflessly given up their time to work for "pride,"
have to integrate this critical volunteer position with their real jobs,
which pay the bills and the mortgage on their homes, as well as
provide for their family's other needs? One of them recently told
me that in practice when an emergency call is received, he often has
to leave his family at home, drive to the ambulance center where
the ambulances are stationed, and then drive his ambulance to the

scene of the emergency. Furthermore, it is not infrequent that one ambulance crew in a particular geographical location may not be manned at a certain time or cannot take the call, so if the emergency call that has been dispatched to a particular station by the call center is not answered within a certain period of time, the telephone dispatch office will automatically send it to the next ambulance service in another geographical location and so on until it is finally answered, all the time creating many minutes of delay. Yet in terms of response times, the statistic that is recorded is the time taken for the ambulance service that actually received and took the call from the dispatch center to arrive at the scene of the emergency, not the time taken for the ambulance to arrive after a distress call was first made to the Suffolk County telephone center.

Not to slight any of the companies, but it is clearly more difficult to have a uniform system for dealing with cardiac arrests with so many ambulance services operating in one county and on a voluntary basis. In the United States there are more than three thousand counties spread across the different states, and most of them (like Suffolk County) rely on volunteers working through different ambulance companies to provide a large component of their emergency responses. It is hard to believe that, in the most affluent country in the world, we can't establish a unified service where we pay all our emergency responders. We wouldn't expect our doctors or nurses to work on a volunteer basis, nor would we do so for the people working in an airport control tower. Yet we do with our emergency responders.

The FDNY in New York City, which has seven times as many residents as Suffolk County and is only about forty miles away, has demonstrated that a centralized, nonvolunteer system works better. In New York City, for example, ambulance crews now cool patients immediately to slow down cell damage, and as Dr. John Freese, who pioneered this system, reported, this has improved patient outcomes. The FDNY is constantly working to set new standards for lifesaving measures. It also insists on certain protocols from hospitals so the crews' work is not for naught once they bring the patient to the emergency department. London also has an efficient single system, as does Paris, where in my mind the system is even better because ambulance crews attending cardiac arrests also have an emergency doctor with them, thus giving a higher level of immediate care and expertise to patients.

The course to improving out-of-hospital CPR rates is fairly straightforward. There must be an increased rate of bystander CPR (which is CPR given by ordinary citizens who witness a cardiac arrest prior to the arrival of EMS personnel), so that CPR can be started immediately rather than after the paramedics arrive. This requires more training within the community, like that which has been done in Seattle. Having defibrillators in the community is also important—yet another initiative that began in Seattle. Ambulances must arrive promptly, in less than five minutes, and the ambulances must have personnel who can deliver a high quality of chest compressions, whether done manually or through the use of portable chest compression machines, which can do the job more uniformly and consistently without the problems of human error or fatigue that have been highlighted by many research studies, such as those by Drs. Dana Edelson and Benjamin Abella. People can help their communities by becoming trained in basic life support. Finding a course locally in the United States is relatively easy and can be done by contacting the American Heart Association (www.heart.org) or the American Red Cross (www.redcross.org). Another excellent resource is the Citizen CPR Foundation (www.citizencpr.org), which holds regular conferences and has excellent links to other organizations involved with resuscitation.

Improving these basic areas can lead to survival rates increasing from 0 percent to as high as 21 percent in the community. In fact, as Dr. Graham Nichol at the University of Washington in Seattle concludes in one of his studies, if all communities in the United States

used the Seattle model, we would have fifteen thousand more cardiac arrest survivors every year. That's a staggeringly high number
on an annual basis. Over ten years that would be 150,000 people. I
actually believe we could save a lot more, since this only focuses on
out-of-hospital cardiac arrests (i.e., where people died and attempts
at resuscitation had been made only up to the point where a person
is delivered to the hospital). There are also many more cases in
hospitals. Optimizing and standardizing hospital care is one of the
major areas where improvement is also needed urgently and where
a huge impact can be made in survival and quality of life.

PROGRESS HAS BEEN MADE in some places. In New York City, knowing that cooling is now best practice, the FDNY concluded that it
is not safe to take the patients to hospitals that did not offer that
treatment. Consequently, the FDNY mandated that as of January 1,
2009, every hospital that receives cardiac arrest patients must have
a designated hypothermia program. This action led to New York
City hospitals finally uniformly adopting cooling (this is an example
of how external factors can influence an organization). Prior to this,
cooling was piecemeal—some hospitals had it and some didn't—
but even after this initiative, within individual hospitals some units
cooled patients but others did not. This is because even some doctors familiar with the landmark 2002 studies remained skeptical of
cooling patients in the hospital and nobody enforced it.

The studies clearly showed that for every six people treated by
cooling, one more person had a benefit. This adds up to an enormous
number of people. If you consider a sample of 350,000 people (which
is estimated to be the total number of cardiac arrests occurring in the
community in the United States each year), then you would potentially get up to fifty thousand more survivors.* However, many doc-

---

*In reality, it would probably be less, since some of the 350,000 cases would include
patients with intractable conditions such as advanced cancer.

tors decided not to provide the treatment to patients in the hospital because the actual studies were conducted on patients who had suffered a cardiac arrest out of the hospital. And, as a practical matter, it would now be unethical to design a study for in-hospital patients suffering with cardiac arrest and withhold cooling from half the participants, so there will never be such a study since the benefits are clear.

The same dilemma exists in pediatrics. Cooling studies have not been specifically conducted on children because it is generally tricky to perform research studies on underage children. As a consequence, some doctors will withhold cooling from children in cardiac arrest cases because they say there haven't been studies specifically in children and thus they don't have enough data to support this treatment for children. But again, to me and many of my colleagues, their refusal doesn't make sense if one understands the rationale for the use of cooling. Even though specific studies haven't been done in children, numerous studies have shown benefit in animals, neonates, and adults. The brain and other organs of neonates and pediatric patients are not significantly different. At my hospital, we had a case where two kids, one nineteen and the other sixteen, fell into a cesspool and were overcome by toxic gases. It took twenty minutes for paramedics to rescue them. When the boys arrived at the hospital, they were both in terrible shape, having had all the complications of a cardiac arrest. If we were to follow things didactically, then by definition the sixteen-year-old should not have received hypothermia, but the nineteen-year-old should have. Again, this is arbitrary and doesn't make sense. As already pointed out, aside from hypothermia, many other important items need to be incorporated during the postresuscitation phase, including providing early cardiac catheterization; optimizing blood pressure regulation (these patients often require higher than usual blood pressure to get blood into the brain); preventing seizures, since it is estimated that around a quarter of patients develop seizures that lead to long-term brain damage; keeping oxygen levels on the relatively low side of normal, since excess oxygen itself is toxic to the brain; and ensuring the levels of carbon dioxide in the blood are normal (otherwise it affects brain blood flow, which if too little or too high causes more brain damage).

IN 2009, THE AMERICAN Heart Association sponsored a conference called the Cardiac Arrest Survival Summit and released consensus recommendations in a report titled "Implementation Strategies for Improving Survival After Out-of-Hospital Cardiac Arrest" to address the disparities and variation in care across the United States for people who suffer from cardiac arrest and the difficulties in implementing optimal standards of care for everyone. The goals were to take the latest discoveries and best practices discovered through research and "translate" them into common practice, to make sure they were understood, and to determine how to implement them. The conference featured many of the leaders in the field of resuscitation science, like conference chair Dr. Robert Neumar, and equally as important, it gathered representatives from multiple disciplines involved in all stages of cardiac arrest care. In addition to the physicians in critical care medicine and resuscitation science, the conference included insurance company representatives, officials from regulatory agencies, ambulance personnel, nurses, funding agencies, research scientists, and people from the general public. It was a concerted effort to involve everyone who has anything to do with cardiac arrest, because the only way to change outcomes is to bring together all those with a stake.

The conference focused on the central problem of why survival rates vary so much across the United States and around the world and why more people aren't being revived, kept alive, and sent home without brain damage, like Joe Tiralosi. Participants zeroed in on many of the critical issues. "The differences in outcome after cardiac arrest do not appear to be fully explained by differences in patient characteristics," the conference's report stated. "Rather, the high rate of survival observed in some communities suggests that

OHCA [out-of-hospital cardiac arrest] is a treatable condition and that outcomes may depend on the effectiveness of the system of care. Ongoing comprehensive surveillance of OHCA events and outcomes through hospital discharge is necessary to identify opportunities for improvement so that all communities can achieve higher rates of survival. The absence of a national surveillance system is a barrier to such an effort, and available resources are insufficient to support it on an ongoing basis."

THE PROBLEM LIES NOT just with cardiac arrest but also with much of medicine. The issue is called knowledge translation, which is essentially putting research into action. The knowledge translation process is dependent on everyone, from those who fund research to doctors who treat patients to the general public. However, since the roles of those involved are not clearly defined, studies have shown that knowledge translation is often haphazard, slow, and unpredictable even in the best-funded fields of medicine—making it all the worse for an "orphan" like cardiac arrest.

Clearly, steps can be taken to raise survival rates if these can be implemented on a wide-scale basis. As difficult as this may be, it can be accomplished. Currently, there is a body called the International Liaison Committee on Resuscitation (ILCOR) that is composed of experts from around the world. ILCOR has more than two hundred doctors with expertise in cardiac arrest, who painstakingly review all the data that has been published in the research literature on cardiac arrest. They meet every five years and debate international guidelines for cardiac arrest. Once these guidelines are accepted by ILCOR, they are endorsed, published, and disseminated by numerous national and international bodies such as the American Heart Association in the United States, the European Resuscitation Council in Europe, as well as the Australian and New Zealand Council on Resuscitation, the Heart and Stroke Foundation of Canada, the InterAmerican Heart Foundation, the Resuscitation Council of Asia, and the Resuscitation Council of Southern Africa. However, if doctors do not know where the guidelines are, then they obviously aren't going to be very helpful. Either way, these are guidelines, not mandated requirements, and as such there is no enforcement body to ensure they are followed. In fact, few or maybe even no hospital that I know of fully incorporates all the recommendations and links in the precardiac arrest, intracardiac arrest, and postcardiac arrest (postresuscitation) phases at all times. As the AHA conference report stated: "The Institute of Medicine has recognized that emergency medicine lacks a standard set of measures to assess the performance of the full emergency and trauma care system in all communities, as well as the ability to benchmark that performance against statewide and national performance metrics. In this respect, cardiac arrest is similar to other acute life-threatening illnesses. You must have benchmarks so you know you are approaching the standard."

IN THE UNITED STATES, an organization called the Joint Commission for Health sets quality standards for medical care in hospitals. The commission looks at certain conditions and declares them measureable entities, such as the rates of infections in intensive care units in hospitals, but unfortunately they don't cover every important medical condition or complication. Cardiac arrest has traditionally not been covered by the Joint Commission as a measurable entity, so hospitals have not had a mandatory need to comply with any recommendations issued by the AHA or any other body. Thankfully, in 2010 the Joint Commission did finally set up an initiative to make core aspects of cardiac arrest performance in hospitals measurable entities. However, instead of simply working to adopt and, most important, implement all the recommendations set up by ILCOR, which have been endorsed and disseminated in the United States by the AHA in separate guidelines published in 2008 (focusing on postresuscitation treatment) and again through a new set of guidelines published in 2010 (regarding the whole spectrum of precardiac arrest, intracardiac arrest, and postcardiac arrest care), the commission worked on a small number of items. In fact, the Joint Commission started with nine items that members considered important, which included the use of hypothermia but sadly only mandated its use for adults who have suffered a cardiac arrest in the community and not in the hospital setting. It also excluded children and neonates.

So hypothetically speaking, based on this standard, if a thirty-five-year-old man's heart stops inside a hospital building, there is no mandatory requirement for the hospital to provide hypothermia treatment, but if the same man's heart stops in the street outside the hospital, there will be a requirement to do so. Or if a seventeen-year-old person's heart stops, there is no requirement to provide hypothermia whether in the community or in the hospital, but if the same person's heart stops on the streets outside the hospital a day after he or she has reached eighteen, then the hospital will be mandated to provide hypothermia—but not if the heart actually stopped when the person arrived in the hospital emergency room. The nine items also still missed some of the most important and fundamental recommendations, such as the time to the start of chest compressions and, importantly, the quality of chest compressions. As we have seen, there is no point giving compressions if they are not of a high quality, so if hospitals are not mandated to measure the quality, how will they know if they need to improve it or not? The commission also did not include the critical issue of overventilation (breathing) rate, which, as discussed, if in excess leads to death since it prevents the heart from restarting; instead the Joint Commission focused on the insertion of a breathing tube without focusing on how many breaths are given through the tube. Other important areas that are missing are competencies for staff regarding skills in delivering resuscitation care, the use of a specific quality parameter called "end tidal carbon dioxide," which indicates the overall quality of the circulation and hence chest compressions, the time to patients receiving cardiac catheterization, blood pressure regulation, prevention of seizures, the levels of oxygen being administered, as well as the levels of carbon dioxide in the blood. Each and every one of these factors, if ignored, can lead to devastating consequences.

Unfortunately, since the original nine items were introduced, and following a system of discussion, the Joint Commission members have actually reduced the original nine items to four. Initially, actual survival figures were part of the nine items, but they have since been removed. So without a mandatory requirement for reporting of actual survival figures among institutions, how can we strive to improve care? If some institutions are achieving only 15 percent overall survival while others are achieving 30 to 40 percent, it would be difficult to identify areas that need to be rectified. The final four items that have been agreed on are the time to shock treatment of the heart, the insertion of a breathing tube, the initiation of cooling treatment for adults who have suffered cardiac arrest in the community only, and, finally, ensuring the right temperature is maintained for these people. After a pilot phase, the goal is to introduce these items in 2013. Although the fact that the Joint Commission has adopted a minimum protocol is commendable and better than not having one at all, since the protocol items are missing the major weaknesses in the system, it is very unlikely they will make a significant enough difference in outcomes. Unfortunately, policymakers will be left thinking that some external regulation actually exists, but clearly the regulation will at best be severely insufficient; the zip-code lottery of care will just continue. It's like setting up an aviation standard that misses some of the most important safety issues raised by the world's leading authorities and the results of published research studies. Since cardiac arrest is an orphan without an FAA-like monitoring body, many of the main issues that are related to outcomes have not been fully addressed. The bottom line is that there is no national or international standard.

The whole world witnessed the devastation that happens as a result of a cardiac arrest and the potential difference in outcomes, as well as the impact on families, in two similar young people. The first, Fabrice Muamba, a twenty-three-year-old professional soccer player, was playing for Bolton against Tottenham in a live televised FA Cup game in England when in the middle of the game he collapsed. His heart had stopped. He died. The world witnessed the ensuing events in horror on live television. The paramedics attempted CPR on the soccer field. A cardiologist, Dr. Andrew Deaner from the London Chest Hospital, who happened to be watching the game, ran on to the field to help. After about ten minutes without a heartbeat from Muamba, the paramedics decided to transfer him to a local hospital. Dr. Deaner, however, insisted that Muamba be taken to his hospital and his intensive care unit and nowhere else, even though his hospital was farther away. I watched the events on television and was horrified like everyone else, although I wished I could have been on the field. I kept telling my wife, "I hope they do all the right things, and I hope they cool him."

Muamba did not have a heartbeat for almost an hour and a half before it was restarted in Dr. Deaner's hospital. Muamba was given hypothermia treatment, and to everyone's astonishment he recovered and was able to leave the hospital neurologically intact a month later. I couldn't help but wonder why Dr. Deaner insisted that Muamba be taken only to his hospital. He was probably uncertain of the care that Muamba would have received elsewhere, and at least he felt secure that he could help bring better care in his own unit. If Muamba had gone elsewhere or with a different group of doctors, they may have stopped trying to get his heart started well before the hour and a half that it took. Traditionally, people don't go beyond ten or twenty minutes, and in the case of a young person, sixty minutes would be considered a long time. However, they did continue trying and with appropriate postresuscitation care delivered him back to his family and to the whole world.

Almost four weeks later, on April 14, 2012, just as Muamba was preparing to leave the hospital, the exact same thing happened in Italy. A twenty-five-year-old professional soccer player, Piermario Morosini, was playing for Livorno in a game against Pescara when in the thirty-first minute he collapsed. His heart, like Muamba's, suddenly stopped beating, and he died in full view of the television cameras. The other players were devastated and were visibly crying while paramedics attempted CPR. There was a report that there may have been a delay getting an ambulance to the scene, as it had been blocked out from the entrance to the stadium by parked police cars. Nevertheless, Morosini received some CPR on the field before being taken to a local hospital once the ambulance arrived. However, he was declared dead soon after. Dr. Leonardo Paloscia, the cardiologist at the hospital where Morosini was taken, said, "Nothing could be done. It was all ineffective, after about an hour and a half of intensive care, his heart never made a beat."

Listening, I found it hard to know what he meant by "nothing." Had Morosini received the quality of CPR needed for an hour and a half? Studies have shown that most people cannot deliver the required quality. How about the breathing rate—had that been maintained? There are many more details that would make the difference, and clearly I don't know all the details, but what was most obvious was that if this young man had been taken to the hospital in Japan where the young woman with an overdose had been taken, he would have been placed on an ECMO machine. This is a form of heart-lung bypass, and it can artificially provide the oxygen and circulation required to maintain his organs even when the heart doesn't beat, while giving the doctors much more time to find the problem that had caused him to die and so fix the underlying problem. Then with good postresuscitation care, as with Muamba, he likely would have escaped brain damage and potentially would have even played again. Again, it is hard to know, but the reality is the team in Italy most likely did the best that they could under the circumstances, and this is why we need a system of care to be developed. If ECMO is helpful in Japan and South Korea, then it is helpful in Italy too—in the same way that if radar is helpful in Japan and South Korea, then it is helpful in Italy too. We don't have these disparities in aviation, but we do with cardiac arrest resuscitation. This is the difference between one person coming back to life and another not.*

Unfortunately, the saddest part of the whole story was that Morosini had a disabled sister who had relied entirely on this young man for her financial and emotional support since their parents had died seven years earlier. They had one other brother and he too had died, so she had only Morosini to rely on. After he died, she had no one to care for her. It is therefore not just the story of Morosini's death but also of the impact on his sister. This is why we need to develop the same standards of care for everyone, so that you and your mother, father, brother, sister, spouse, or children don't end up receiving a type of zip-code lottery care, and this is why we need to continue to put aside old perceptions and implement the highest standards while continuing to study through science what happens when we die. Let's not forget that cardiac arrest will happen to every one of us. It is inevitable.

Until hospitals implement reforms and create a uniform checklist to perform, survival outcomes will not increase. The AHA conference report concluded: "Organization of the system of care appears to have a larger effect on survival than patient factors. The creation and maintenance of an effective system for delivering optimal emergency medical care are complex. Examining either systems with historically good outcomes or systems in which change has improved outcomes provides an opportunity to identify best practices that can be broadly implemented." In other words, the problem

---

In fact, the disparities exist even in the same community in each nation.

is not that the patients who have a cardiac arrest are very sick and probably won't make it. If there is a system in place, survival rates will increase.

In spite of the need to improve our systems and implementation techniques across the board, the fact that we can now reverse death begs a fascinating question. Since we now know that cells, including brain cells, that are not functioning can still remain viable (in the sense that if supplied with oxygen and nutrients they can gain function again) for a period of hours after death, and that death itself, medically speaking, is a global stroke that affects the entire brain (also termed anoxic brain injury) and leads a person to go into a deep coma within a few seconds, what actually happens to our mind and consciousness (that entity the Greeks called the "psyche" or "soul"), or put more simply, our "real self"? Does it become annihilated immediately after death, or does it continue to exist for a period of time after death? And if so, for how long?

CHAPTER 6

What It's Like to Die

W<small>HAT IS IT LIKE</small> to go beyond death? What happens in the period after we die but before all the cells in our body have become irreversibly damaged and have reached a point where they can no longer support life? If we could go beyond this threshold, what might we tell others about this experience, and what would they think of us? Would they even believe us?

In the opening sequence of Clint Eastwood's movie *Hereafter,* a woman dies in dramatic fashion, and we get a visceral sense of what it might be like to die. The sequence opens with cuts between a young girl and a woman strolling casually through a local market in a coastal town in Thailand and the woman's boyfriend waking up in their beachside hotel room. An eerie feeling is cast over these normal events when the boyfriend gazes out over the water and sees a huge wave forming in the distance. Suddenly, there is a huge roar and a tsunami engulfs the resort. As the torrent of water careens through the streets, destroying everything in its path, the woman and the girl run for cover, but it's no use. They are both swallowed by the massive wave and carried away with the debris.

As the woman drowns, she fights for her last breath. It's horrifying to watch—until the point when she actually dies. Visions of people swirling in an arena of comforting light begin to form, and an image of the woman and the girl she was shopping with comes into focus. The woman is clearly undergoing a positive and transformative experience. After the tsunami has subsided, the woman's limp body is pulled from the water. The paramedics perform CPR on her, but there appears to be no hope to resuscitate, so they move on to the next victim. Seconds later, the woman begins breathing and comes back to life. Though she was clearly dead, she reported seeing images from her life. She has experienced a vision of the experiences that may take place after the process of death has started.

The fact that Clint Eastwood tackles head-on in a major studio movie the question of what happens after we die shows that this experience has become part of mainstream culture and thought. Eastwood provides no exposition on the scene, and none is needed. From the powerful and moving images, the audience can experience what it is like to die by seeing the woman have a reaffirming vision of light before being brought back to life. As the film unfolds, it shows how her life has been positively transformed by the experience.

People have now regularly died, come back to life, and told us what they have experienced in that early period of death. Their experiences have been fairly uniform and share many common factors. Like the scene in *Hereafter,* people have generally described the experience as being a transformative and positive one. Their lives have been enhanced by the experience, and as a result they describe themselves as being more altruistic and less materialistic, less self-centered and less afraid of death.

It is difficult to trace back exactly when we became aware of these experiences, but most likely it had something to do with the discovery of the new field of resuscitation in 1960 and its increasing acceptance by the medical community in the 1960s and 1970s. Following the rise of resuscitation science, by the mid-1970s, there were clearly more and more people who had been successfully resuscitated from death. The interest of the scientific community and the general population in these experiences was heightened in 1975 when Raymond Moody, a psychiatrist and university lecturer in philosophy, published the bestselling book *Life After Life*. Moody's book was the first comprehensive study of the human experience during the dying process. He collected and compared the accounts given by 150 survivors of near-death encounters and termed these experiences "near-death experiences," or NDEs. He defined an NDE as "a clinical situation that would normally have led to the death of the individual without medical intervention."

Moody found that though not everyone experienced the same things, there were many recurring features. The most common experiences people described were feelings of peace, happiness, and joy and of being pain-free. Some described having an out-of-body experience, where they felt they had separated from their physical body and they had looked down on themselves from above. Some experienced an instant panoramic review of their lives, in which they experienced all that they had said and done, while others described an encounter with deceased relatives, who seemed to have come to welcome them. Others recalled seeing a bright light and going through a tunnel and in some cases meeting a luminous being who radiated a feeling of absolute love and compassion. They often felt as if they were entering a beautiful domain, though many also talked about a point beyond which they could not cross, or else they couldn't return back to life. Oftentimes, people said they returned reluctantly to life, because what they experienced was incredibly beautiful. Words could not truly describe the beauty they had witnessed. One particularly noteworthy feature was that many people's attitudes, beliefs, and values were permanently and dramatically altered. They felt a new sense of purpose and appreciation for life. They also became less afraid of death. Much like Joe Tiralosi, those who had become most positively transformed were those who had encountered a being of light.

Today, NDEs are generally understood to be the often vivid and realistic, and sometimes profoundly life-changing, experiences occurring to people who have been physiologically close to death or have even gone beyond the threshold of death, as in cardiac arrest. At the beginning of his book, Moody gives an example of a complete NDE based upon all the common features. He emphasizes that this was not a particular person's experience, but rather a model or composite of the features commonly found in NDE accounts:

> *A man is dying and, as he reaches the point of greatest physical distress, he hears himself pronounced dead by his doctor. He begins to hear an uncomfortable noise, a loud ringing or buzzing, and at the same time feels himself moving very rapidly through a long dark tunnel. After this, he suddenly finds himself outside of his own physical body, but still in the immediate physical environment, and he sees his own body from a distance, as though he is a spectator. He watches the resuscitation attempt from his unusual vantage point and is in a state of emotional upheaval.*
>
> *After a while, he collects himself and becomes more accustomed to his odd condition. He notices that he still has a "body," but one of a very different nature and with very different powers from the physical body he has left behind. Soon other things begin to happen. Others come to meet and to help him. He glimpses the spirits of relatives and friends who have already died, and a loving warm spirit of a kind he has never encountered before—a being of light—appears before him. This being asks him a question, nonverbally, to make him evaluate his life and helps him by showing him a panoramic, instantaneous playback of the major events of his life. At some point he finds himself*

approaching some sort of barrier or border, apparently representing the limit between earthly life and the next life. Yet he finds that he must go back to the earth, that the time for his death has not yet come. At this point he resists, for by now he is taken up with his experiences in the afterlife and does not want to return. He is overwhelmed by intense feelings of joy, love, and peace. Despite his attitude, though, he somehow reunites with his physical body and lives.

Later he tries to tell others, but he has trouble doing so. In the first place, he can find no human words adequate to describe these unearthly episodes. He also finds that others scoff, so he stops telling other people. Still, the experience affects his life profoundly, especially his views about death and its relationship to life.

Moody observed that most of the people who had an NDE did not experience all the same features and that the sequences in which the events took place varied. Further, he noted that whereas some people remembered only two or three features of the NDE, others recalled more detailed experiences.

After the publication of *Life After Life,* NDEs became the subject of great controversy and heated debate. Although many people claimed that these experiences provided a glimpse of the afterlife, others in the scientific community were more cautious, claiming that they were at best hallucinations and at worst fabrications owing to the publicity being generated by the book. But when researchers began to study these accounts in detail, they found very little evidence to support this view, as many of the accounts predated the 1970s and the publication of Moody's book. In fact, some of the accounts went back decades, and in the vast majority of cases, the people involved had consistently declined any publicity and had only shared their experiences with researchers. Often they had not even mentioned them to their friends and family. Interestingly, then, although records of cognitive experiences from a

close brush with death were found that could be traced back millennia, these experiences became much more clearly described and
hence more popular in society from the 1970s onward.

When I began my own research into NDEs in 1997, there was
a lot of debate that complicated the mystery even further. This was
mainly because the descriptions and feelings that people had experienced when examined individually could also occur in people
who had not had a close brush with death. For instance, it was not
only those who had come close to death, or had even gone beyond
it, who would see a bright light, feel peaceful, or have a transformative experience. The other confounding factor was that the term
near death was extremely ambiguous and vague; as with anything in
science that is poorly defined, it had created a lot of discussion and
also disagreement. Nevertheless, the public and scientific interest
had led to a comprehensive historical and scientific examination
of these experiences and the factors that may cause them, stirring a
spirited debate that encompassed science, psychology, religion, and
even culture.

After I started my work, I received hundreds of letters from
people who claimed to have had an NDE. I began to study the
claims of those people who satisfied the criteria laid out by Moody
in *Life After Life,* and then I delved further into their cases. Adults
had a wide range of experiences, and reviewing these was critical to
finding a scientific process to study and interpret NDEs of patients
whose medical history and treatment through the time of death was
known. Evaluating and comparing the experiences of the broadest
possible cross section of patients would be a step toward learning
what happens to people who die or at least have a close brush with
death and are then brought back to life.

One of the most consistent features described was a sense of
calm and peace that lasted throughout the experience. One woman
who suffered a gynecological hemorrhage said that she was surrounded by a strong light and felt happy, peaceful, and not the least

bit frightened. She recalled that she was "high on the ceiling of the ward looking down upon the bed (which seemed to be a long way down) and saw the doctors and nurses around the bed working on the person lying there." In a layperson's terms, she had an out-of-body experience.

This feeling of having an out-of-body experience was prevalent in people who had NDEs. A majority of the people described the sensation of separating from their bodies and being able to see events happening below while floating in an out-of-body state. They likened this to removing a heavy garment of clothing or shedding a skin and moving away freely, leaving the old skin behind. Interestingly, people consistently described their "self" as the part that was above, rather than the body that was lying below (suggesting they associated the self with the entity that they felt had separated). In other words, an individual would say, "I was at the corner of the ceiling, looking down at my body." Almost uniformly people described feeling very peaceful and not at all concerned that they were watching themselves die.

In the most complex cases of out-of-body experiences, people were able to recall the details of what was happening to their bodies during the experience, largely because most claimed to have seen it from above. One woman recounted floating above the bed watching the doctors work on her and also listening to their conversations. After she survived the operation, the surgeon visited her and told her about the resuscitation procedure. The woman said that she already knew the details because she had heard him talking about them during the operation. The doctor dismissed her recollection, saying that she must have heard a nurse discussing it after she had regained consciousness.

In addition to recalling specific details and describing the self as the part that had detached from the body, this woman's case also illustrated a common reaction of disbelief from the medical and nursing staff when told by the patient what they were doing during the resuscitation process. The medical staff often rejected the idea that the patient could have seen anything because the person was unconscious. Nevertheless, there were many similar cases in which people who were seemingly medically dead correctly described the events going on around them.

Later I also found many other cases with very similar features to those I had collected in the United Kingdom. For instance, in the United States one woman named Lauralynn, a trainee professional tennis player, reported that she had been scheduled for a routine operation, which her doctors had told her would take twenty minutes. Unfortunately, during the surgery, the doctors punctured her abdominal aorta. She suffered extreme blood loss and had a cardiac arrest and died. At that moment, she found herself near the ceiling, where she could see her body on the table. From there, she watched the surgery team frantically working on her body. She remembers people being very upset, though she wasn't sure what was wrong. She said that she had no pain or fear; she knew everything was going to be okay.

Lauralynn then described how she found herself in a place of total blackness. She described this state as "comfortable, peaceful, and quiet" and recalled seeing a bright warm light on the horizon that beckoned her. She wanted to go into that light. "It was a place of unconditional love; it was a place that I would never want to leave."

But Lauralynn also described meeting her brother-in-law, who had died seven months earlier from cancer. He took her on a journey that made her understand her life. After the journey was finished, he told her it was time to go back. Though she didn't want to return, he insisted. He told her: "You have to go back. You have to live your life's purpose." With that, Lauralynn says she was pulled back into her body.

Her life, however, was completely changed. She said that after the experience she felt "like a stranger in a strange land." She also felt disoriented and couldn't grasp why everyone was running from place to place and task to task. The biggest change was that she learned that life was about giving yourself to people. On a day-to-day basis, she said she began trying to live each day to the fullest as though there may not be another one.

One of the most consistent features that people had during NDEs was seeing a tunnel, though it was described in different ways. Some described it as a long, dark tunnel; others said that it looked more like a kaleidoscope with colors on the sides. One person described the tunnel as "dark on the top and bottom, but not solid. The sides were like tiles—some were red, yellow and green, and the others were black. All had a shine to them." After studying many cases of those who recounted seeing a tunnel, it became clear to me that everyone was basically describing the same thing. To underscore this point, often people who recalled the tunnel saw a bright light at the end, which they interpreted as warm and welcoming.

Besides the bright light, people mentioned seeing an actual luminous being, a being of light. The being was described by both religious and nonreligious people as containing love, mercy, and compassion, and in their eyes, this being was absolutely perfect. Some people identified the being of light as God; some thought it to be a religious figure such as Jesus, and others interpreted it as a simple, nondenominational being of light. For some, this was someone who played the role of a loving educator, who exuded compassion and watched over the person and guided the individual through the experience and a review of his or her life. In the review, people found themselves seeing and experiencing everything they had done in life and felt that their actions were completely transparent. The prevailing feature during the whole experience was a deep sense of all-pervading love and benevolence. The love that emanated from the being of light had a far deeper intensity than that which emanated from other people (such as deceased family members) who were encountered during the NDE or anything they had encountered in life up until that point. Some also described feeling the pain and distress that they had caused others whether intentionally or unintentionally during their lifetime. After surviving, many made a promise to themselves that they would try not to harm others with words or actions, and this led them to view life as an opportunity to be a source of goodness to others irrespective of creed or culture.

Some people who saw the tunnel went through it. The description of this sensation ranged from individuals moving forward with their own power through the tunnel to individuals who said they were pulled through it by an unknown force. Several people who went through the tunnel recalled arriving in a beautiful, gardenlike place. In many cases, the sensation of going through the tunnel was associated with other NDE features, such as encountering a being of light or seeing deceased relatives at the end of the tunnel. One person who went into anaphylactic shock and whose response was typical of this group wrote: "I 'died' and saw all the old relatives who have passed on. There was a large tunnel and all these people were calling me to join them. I started to go towards them."

Although some people described hearing a voice or seeing a person telling them not to go any farther or to turn back, others said they themselves chose not to go farther. For the overwhelming majority, the driving force that made them turn back was the need to care for others, particularly their children. One woman, who described hovering over her bed and watching the doctors, wrote: "To the right of the room up high I sensed a tunnel of light, but I didn't want to go through it, as I had just had my little girl and wanted to go back and look after her, and my husband and parents. I remember drifting back down into my body and back, oh, painfully back." Others described reaching a similar symbolic point of no return, such as a wooden gate, a stream, or a river that they knew not to go beyond because returning would be impossible.

Memory was a significant part of the experiences. After a significant insult to the brain, such as a head injury or a seizure, or a change in blood oxygen, carbon dioxide, or glucose levels, there is normally a period of memory loss before and after the trauma. This can be as short as a few minutes and as long as days or even weeks, and it is due to an imbalance in the normal biology required for the brain to function. The extent of memory loss depends on the severity of the brain trauma. This may have accounted for why some people had varying recollections of the experience and the events leading up to it.

In one case, a traffic accident victim recalled being pulled into a tunnel and resisting, but he could not remember the accident happening or being treated in the hospital. Despite the memory loss surrounding the event, the man had complete recall of the NDE. Joe Tiralosi, however, did not remember any of the specifics of what happened during the forty-five minutes he was medically dead. He did recall encountering an incredibly warm, loving being who had comforted him and made him unafraid of death. He also didn't remember going through a tunnel or seeing a heavenly place, though interestingly he was admittedly left with a life-affirming feeling from the experience.

The most complete NDE case of the initial group I studied came from a woman who gave a detailed account of meeting a "perfect being" who guided her through the experience. When I interviewed her, she said that she found it difficult to convey the feelings of deep compassion, love, and kindness that had emanated from the being. The woman, who was suffering from an ectopic pregnancy that caused massive internal bleeding, collapsed after calling a friend for help. She related:

> *I suddenly found myself standing beside myself looking at a cord which connected me to my body and thinking how thin and wispy it was. Someone was beside me. I was made to feel secure and encouraged to trust my companion, who suggested that the cord was insignificant and that I should not concern myself with its fragility. I was guided*

*towards the light. This was a sort of void, in which I found myself
with the ability to fly, or should I say I had no weight—a very strange
experience. Throughout the journey I kept looking back to ensure my
companion was with me but somehow towards the end of the journey I
found myself just content to move on and reach the end.*

*Reaching the light, I was met by other beings of light and very
gently encouraged to move on towards a life review. In this experience
my actions were not judged by others, I judged myself. My presence could see into my mind and there was no way I could hide any
thoughts. Gently I was encouraged to understand how my mistakes
hurt others by experiencing what others had felt as a result of my actions. I was confused, as it all seemed so strange. The word "death"
was never mentioned yet somehow I came to understand that I was in
that place of spirit where the newly dead move on to.*

*Many questions sprang to mind like how, why? I had just had
abdominal pain, nothing considered life threatening. I was told by those
in spirit that I had been pregnant. I had not known that I had been
pregnant before this; I had just thought that I had had abdominal pain. I
was also told that the spirit of the child had initially consented to be born
and then changed its mind . . . that it had experienced a very traumatic
life before and just could not face life again just yet. Perhaps with love
and encouragement it would in the future. I asked to see this spirit and
explain that with me and my husband it would have known love. We
had been hoping for another baby for some time. There was hesitancy
and one can only assume the spirit concerned was consulted. After a
delay we spoke together. Poor soul, I really sensed the fear. It felt secure
with the brothers of light around it, who supported it with love. "One
day" was the message from the brothers. "Be patient with him."*

*I was moved forward and eventually met the great God in many
religions, a beautiful experience and I can only say that I fully understand why St. Paul so wished to be with him, to be in the presence of
such unconditional love, humor . . . understanding . . . I did not need
to speak—thoughts were sufficient. It was as if all were one and shared*

in his being; his radiance was everywhere. To this day I still look back
with elation at this experience.

I became very distressed and became very concerned at leaving
a young 18-month-old baby behind. Who would care for her? My
husband was away, no family close by. [God's] compassion was so
strong, his love and caring so abundant that by his grace I was allowed
to return. I was told that I would have a very special mission to do
later in life, when my children grew up. He already knew there would
be another.

I cannot remember much about the return. I recall being at the
ceiling of the room and watching two nurses either side of me working
on drips and drains. There was a jolt and in no time I had drifted into
what can only be described as sleep.

I had never read of near death or out of body experiences, terms
used today. Two years later my son was born, very sickly, but to this
day I have kept my promise given to him in that world of spirit—that
I would love him unconditionally as long as he needed me. I created a
home of love and as a family we work together to love each other and
the world—a small mirror of what I experienced in that land of light.
After this experience I have no fear of death and believe with certainty
in the afterlife.

The most outstanding feature of this case was the woman's in-
teraction with the being of light and the transformation it had on
her life. Although many years had passed since her experience, it
was still fresh in her memory. When I met her, she told me that
during the process of judgment she felt uncomfortable and remorse-
ful about the opportunities she had failed to make use of in her life.
She described these as situations where she could have had a positive
effect on others but did not follow through. She also told me that
she now hated to do any harm to others, as she had felt the pain that
she had caused others. She now felt that the most important thing in
life was to take the opportunities to be of assistance to others, even

if these were sometimes the more difficult options. This response also highlights the positive transformation that many people undergo following an NDE, particularly those who have encountered a being of light.

In many ways, this case shared many similarities with the case of an American man named Steve whose complex NDE I came across many years later. The experience recalled by this woman as well as what Steve had experienced provided some real insights regarding what it is like to undergo a life review. Steve had explained that after having an asthma attack that had caused him to die, he experienced something unusual. He explained, "Everything was surrounded by this light blue grey color. I found there was a being beside me. I could feel his presence. It was a comforting presence, a reassuring presence but was also a presence of magnitude and power. I felt things were alright. Then I began a review of my life, of the key moments of my life. But at the same time I was re-experiencing it from the other people's points of view and that was a stunner because you feel their pain, you feel the sting; you feel the hurt [caused to others by your actions]. It was a horrifying realization that I wasn't the person I thought I was, and at that point the being then sent me messages to explain 'it was alright, that was what humans do, humans make mistakes.'" What was most interesting was that for this man, much like the British lady, the life review entailed reexperiencing past events from other people's perspectives. In essence, Steve had felt and experienced firsthand whatever pain or discomfort he may have caused other people in life, as if someone were doing the exact same thing to him. Thus his own being had become the judge of his own actions and behavior toward other people, without the need for external judgment.

Everything was obvious and clear. With parts of his life laid bare, Steve was shocked at how poorly he had treated people. At times, he had been deceitful, had hurt people, and had outright lied—all of which he had justified to himself at that time, rationalizing that people deserved what they had coming. But during the experience, he felt the other people's pain caused by his actions. Being forced to relive these events made him question his prior perspective on life and led him to conclude that living a life in that manner had been a failure; he felt that what he had done had been "humiliating and dreadful."

Steve returned to consciousness in great pain. He was paralyzed from the neck down and was told his condition was not reversible, though thankfully ultimately he recovered and was able to regain full motion. He later described the reason he hadn't liked what he had seen during his experience, which was due mostly to what he had learned from the astonishing life review he had undergone. He explained, "I re-experienced these events [referring to the actions he had performed in life] from my own point of view. I wasn't just watching the events; I was actually reliving them again, while at the same time I was also re-experiencing the actions from other people's points of view. I was them. I was reliving the experience from their point of view and at the same time (and I don't know how this works) I was also experiencing it from a higher reality; the truth of the matter. So what I saw was my own lies and my own self deception to myself, which I had used to convince me that doing certain things was okay because people had deserved it. Then I was experiencing the emotional impact it had on other people. I felt their pain. I felt the shock on them. But then at the same time I also saw that they have their own lies and self deceptions and so the net result was that I felt like I was a failure as a person and I wasn't the person I had thought I was. It was humiliating. I felt really dreadful and it was completely humbling." He emphasized that "the judgment came all from myself [within his own being]. It was not from an outside source, but then this being that was with me was also sending me comforting messages—thank goodness!—and one of them was it was alright as I was only human." The entire experience, he said, made him feel as though he was being given a second chance to live a more meaningful life. He felt he had a chance now to change things so that "next time I get back to the life review it wouldn't be the same, or at least they would say, "He tried.'" After his experience, he had become a caring, truthful, and positive person. In short, he wanted to make sure that next time he dies and his life is reviewed that he will, he only half jokingly added, receive a higher grade!

Another experience that I came across later was the impact of a patient's NDE on his own doctor, the well-known U.S. cardiotho-racic surgeon Dr. Mehmet Oz, who practiced surgery at Columbia Medical Center in New York and later set up his own television show. The patient, named George, needed heart surgery, though his surgeon, Dr. Oz, was reluctant because he thought his patient would likely die from the surgery itself. Nevertheless George accepted the risks and convinced him to go ahead with the surgery. During the procedure, complications arose, and George began hemorrhaging heavily to the point that his heart actually stopped and he had a cardiac arrest and died. Dr. Oz and his team worked frenetically to try to save the man and attempted to stop the bleeding and also to restart his heart. With great difficulty they succeeded and George was eventually admitted to the ICU. However, for the next three days, he continued to hemorrhage and he remained perilously close to death.

But eventually the bleeding stopped and he started to recover. One day, Dr. Oz was called to the ICU after George had regained consciousness and his condition was more stable.

George explained to his surgeon that he felt like he had been in a "deep, dark" space and had looked up to "see this bright light." He explained, "I began to fall away from it. . . . I was getting further and further away, and I knew that if I lost sight of the light I would die."

To save himself, he had gone toward the light. However, he said that he was never scared. He simply made the conscious choice that he was going to reach the light. When he did, he said that he pushed himself up, and woke up. George's experience after his dramatic surgery had made his surgeon, Dr. Mehmet Oz, question what happens when we die and the existence of a form of life after death. This had also led him to invite people like Lauralynn, Steve, George, and me to his program, and this was how I had come across all these people.

Although there was no way to validate this claim or any of the other claims, the NDEs were very real to those who had experienced them. In addition to retaining a memory of the experience, people consistently described being able to think clearly and lucidly with well-structured thought processes. Essentially, during the NDE, people retained largely the same consciousness and personality. That is, the people who had gone through the experience had retained the same level of knowledge and understanding and essentially the exact same "self" during the experience as they had before it, though the experience did often lead to a positive change afterward.* When I have been in social situations discussing with others what happens when we die, I have sometimes heard people say, "When I die, I will find out." But these NDEs suggested that after death we don't suddenly "know everything" or develop a greater cognitive understanding. Our depth of perception remains as it was before dying. That may also account for why the interpretation of the experience depended on people's prior beliefs and understanding.

In determining the role of preconceived ideas about life, death, and what happens when we die, the most fascinating and potentially conclusive group studied was children who had NDEs. I found two intriguing cases that broke through all boundaries. The first case

---

*By this I mean that people don't suddenly develop a higher perceptual knowledge. They experience certain things, and their interpretation depends on the level of knowledge they had acquired before they died.

came to me from the grandmother of a boy named John who had read about the work I was doing and contacted me. When John was not yet three years old, he had suffered a cardiac arrest. His grandmother related that when John's heart stopped, he turned blue. He looked lifeless. The room erupted in commotion as people began pressing on his chest, frantically trying to restart his heart. Soon, the ambulance arrived and rushed him to the hospital. Along the way, paramedics were thankfully successful at restarting his heart. John's family moved on with their lives.

Months later, after John was discharged from the hospital and returned to his normal life, he was playing with his grandmother one afternoon. Off the cuff, he said, "Grandma, when I died, I saw a lady." He had not mentioned this to his parents. But over the course of the next few months, while playing, he continued to talk in profound terms but with a child's vocabulary about his experience. He said, "When I was in the doctor's car, the belt came undone and I was looking down from above." He also said, "When you die, it is not the end . . . a lady came to take me . . . there were also many others, who were getting new clothes, but not me, because I wasn't really dead. I was going to come back."

John's parents noticed that he kept drawing the same picture over and over again. As he got older, it became more complex. In the picture, he drew himself floating above his hospital bed and attached to a balloon by a cord. When asked what the balloon was, he said, "When you die you see a bright lamp and . . . are connected by a cord." There was no mistaking that he was trying, as best he could, to describe a near-death experience that had also included an out-of-body experience. Interestingly, just like the woman with the ectopic pregnancy, he recalled experiencing a sort of cord that connected him to his body.

A few years later, I was presented with another case that was equally captivating. It involved a three-and-a-half-year-old boy named Andrew. He was admitted to the hospital with a heart problem and had to undergo open-heart surgery. About two weeks after the surgery, Andrew started asking his parents when he could go back to "the sunny place with all the flowers and animals." His mother told him that they would go to the park when he was feeling better. He said, "No, I don't mean the park. I mean the sunny place I went to with the lady." When she asked him what lady, he replied, "The lady that floats."

His mother told him that she didn't understand what he meant and apologized that she must have forgotten where this sunny place was. He said: "You didn't take me there. The lady came and got me. She held my hand and we floated up. You were outside when I was having my heart mended. It was okay, the lady looked after me . . . the lady loves me . . . it wasn't scary. Everything was bright and colorful [but] I wanted to come back to see you."

She asked him whether he was asleep, awake, or dreaming when he came back. He replied, "I was awake, but I was up on the ceiling and when I looked down I was lying in a bed with my arms by my sides and doctors were doing something to my chest. Everything was really bright and I floated back down."

About a year after Andrew's operation, he and his mother were watching a TV program that showed a child having heart surgery. When the bypass machine appeared on the screen, Andrew grew very excited and declared, "I had that machine." His mother said she didn't think he was right about that, but Andrew insisted he was. His mother then pointed out that he was asleep during his operation and therefore would not have seen any machines. He said, "I know I was asleep, but I could see it when I was looking down." She questioned him about this, and he replied, "You know, I told you, when I floated up with the lady."

Later, in an unrelated situation, Andrew's mother showed him a photo of her own mother, who had passed away. Andrew said, "That's her. That's the lady." The lady he was referring to was his deceased grandmother.

John's and Andrew's experiences are remarkable in that the details in their stories correspond to what researchers have found in NDEs described by adults. The content of the children's NDEs was simpler than those described by adults, and the children were limited in their ability to explain all that they had experienced, but they nevertheless seemed to have had the same core experience as adults have reported. To many scientists and medical researchers, including me, this underscored that there was uniformity to these experiences that could be scientifically studied.

At the time I began studying NDEs, there was a consistency to the transformative and positive undertone of the experiences described by people irrespective of their religious or cultural background—a factor that not only unified the experiences of everyone, but also placed the NDE squarely in the realm of scientific study. The experience seemed to be universal and an indication of what we experience after crossing over the boundary of death, reflecting the cognitive and mental state that humans experience in the early stage of death. The exceptions in my personal experience were the people who had attempted suicide. In those cases, people who survived described some very unpleasant traumatic and painful experiences that did not match those who had died involuntarily or from natural causes.

I HAVE COLLECTED MORE than five hundred cases of people who have experienced a close encounter with death and been brought back to life, including those of children, down to three years old. Through my own personal connection to patients and a study of the voluminous research conducted by others, I have come to conclude that the term *near-death experience* is scientifically problematic and should be altered when considered in the context of people who have medically crossed over the boundary of death and been brought back to life (i.e., those who have been revived after a cardiac arrest).

The main problem is that the phrase is too vague. A great deal of controversy has come about because the accepted definition of what is considered "near" death is for obvious reasons ambiguous and unclear. What does it mean to say someone is "near" death? Did the person have a heart attack, a stroke, or a severe infection? How about someone who has entered a state of medical shock? The truth is that none of those can really be defined as "near death." In the same way that a plane rapidly descending to the ground and toward an inevitable crash can be rescued by an experienced pilot, we too can take someone who is rapidly descending toward death and hence near to death and divert him or her away from that inevitability. The question should now be: Did the person die or not?

Although those patients can probably be considered "near death," the causes that lead to being in a near-death situation are enormous, just like the causes that may lead an aircraft to be in a "near accident" state. More important, within the territory of being "near death" just as in the territory of being in a "near accident," there is an enormous range of possibilities. By "near accident" we could mean any aircraft that loses control and descends one thousand feet, or we could mean any aircraft that descends to earth but is pulled up seconds before it hits the ground. Clearly, there is an enormous difference between these two situations. The same is also true of the term *near death,* and it is scientifically too vague and imprecise. We could mean anyone whose death was prevented before he or she had "crashed," and doctors are quite good at preventing these "crashes" from taking place. This is our day-to-day job in the intensive care unit. There is, in fact, no accurate definition of "near death" in medicine. In science, as soon as something is imprecise and vague it creates an opportunity for disagreement, a wide diversity of opinions and controversy, which makes sense because without adequate precision and a clear definition, people end up discussing things differently. So it makes good sense to disagree!

This is what has partly fueled the discussion and divided opinions, especially by the media, into so-called believers and skeptics.

A new and precise term is needed to accurately define the phenomenon at least with respect to those who have actually died and been
brought back and thus enable appropriate scientific research and
accurate discussion; a precise term would also reduce the amount
of emotion people bring to the subject, since death is an enormously
emotional subject for everyone. It is hard to think rationally when
emotion has taken over a subject and a scientific discussion. The
biology of cardiac arrest and actual death, unlike the multitude of
events that could be vaguely considered in the near-death situation, is very precisely understood. This is similar to how the consequences of an actual aircraft crashing from the sky, such as the
breaking of the aircraft, the formation of debris, and so on, is very
well understood, no matter what initiated the crash. Calling the
experience by a different, more accurate name would also enable
us to look at the evidence again with fresh eyes and more clearly
define the experience. Thus, the main reason I now reject the term
is more profound.

As a physician, I know that, according to current medical understanding, people who have had a cardiac arrest weren't near death.
They were *dead*. Physiologically and experientially, by everything
we currently understand, the sensation of lying unconscious on a
table with no heartbeat and no measurable brain function should be
the same for the patient at any point in the death process. As already
discussed, the physiology in minute one, when resuscitation is still
very much possible, or much later, after death has become permanent and irreversible, is well understood and standardized.

For this reason, when the experience occurs in the circumstances of cardiac arrest and the objective period of death, I think
near-death experience would more accurately be termed the *actual*-
death experience, or ADE. Anyone who dies loses consciousness
with the immediacy of a hammer blow, and electrical activity in
the brain ceases in about ten seconds. Scientifically speaking, people
who lose consciousness under these circumstances, by definition,

should not be able to report highly lucid, detailed, and chrono-logically accurate memories and accounts of the experience. And in fact, the vast majority of patients who undergo any brain injury don't remember anything immediately preceding or following the incident. Yet somehow people who claim these conscious mental processes during the period of clinical death enjoy an inexplicable ability to recall details of which they should be wholly unaware.

For this reason, I think the experience could point the way to a new scientific theory of the mind. Many scientists hold the belief that electrical and chemical firing at the level of the neuron pro-duces all our cognitive processes. But patients whose brains have flatlined after death claim to recall details of what occurred while they lay unconscious on the table. Even more incredibly, they claim their bodies were beneath them, while their awareness and their "self" hovered at the level of the ceiling.

These surprisingly accurate observations seem to occur when patients correctly describe details they could only have seen while out of their bodies. In one case, reported as part of a large scientific study published in the *Lancet,* a prestigious medical journal, Dutch cardiologist Pim van Lommel, who has done extensive research into NDEs, described the account of a patient who had awareness and mental and cognitive function during a cardiac arrest in the hos-pital where he worked. As colleagues began resuscitation efforts, they also set about inserting a breathing tube through the man's mouth. As the man's jaws were parted to permit the introduction of a breathing tube, his upper teeth came loose. He had dentures. The nurse removed them quickly and continued her work. After ninety minutes of steady effort, the man's heartbeat was restored, and he was sufficiently stabilized.

A week later, the man was transferred from intensive care to the cardiac ward, and the nurse who attended him during surgery saw him again for the first time since his recovery. The nurse was entering his room to administer medication and had no intention of revealing that she had been present at his resuscitation. But the patient, incredibly, recognized her. This was surprising enough. He had been unconscious the entire time she was in his room in intensive care—in a coma, in fact, as a result of the lack of blood flow from his stopped heart. But then the patient told the nurse that he had seen her remove his dentures. In all, the patient described the cardiac arrest cart beside his unconscious body, the drawer she had hurriedly tossed his dentures inside, and the small room where he had been resuscitated.

Van Lommel subsequently performed his own investigation, and he was admittedly slack jawed at the patient's inexplicable recall of these details. The man told van Lommel: "I was floating up near the ceiling, and I was trying to let everyone know I was still alive because I was afraid they were going to stop trying to resuscitate me." For a long time, van Lommel did not believe in such experiences, but testimonials like this one helped convince him they were real.

Dr. Mario Beauregard, a Canadian neuroscientist, conducted a study with patients who underwent deep hypothermic circulatory arrest (DHCA) at Hôpital Sacré-Coeur, a research hospital affiliated with the Université de Montréal, between 2008 and 2010. These were patients who had been cooled down to an incredible 18°C (64.5°F) from 37°C (98.5°F), a point at which the brain stops working, and doctors can stop the circulation without causing permanent damage. The body is so cold that the cells have such little metabolic activity that even though they are starved of oxygen and there is no blood flow or circulation, they don't become damaged. This gives surgeons time to safely operate on patients without any circulation. Biologically this is the same as what happens to someone who has died. This surgical procedure was employed in patients to repair aortic defects where they couldn't have been placed on a heart-lung bypass machine. The main objective of Beauregard's study was to estimate the prevalence of conscious mental events during DHCA.

Of the thirty-three patients whose cases were examined, three patients reported conscious mental activity and were interviewed for the study, and one reported having an out-of-body experience.

The patient was a woman who had just given birth and required immediate surgery to replace the ascending aorta. According to Beauregard, the woman did not see or talk to the members of the surgical team, and it was not possible for her to see the machines behind the head section of the operating table as she was wheeled into the operating room. She was given general anesthesia, and her eyes were taped shut. Still, she claimed to have had an out-of-body experience (OBE) at one point during the surgery. From a vantage point outside her body, she described seeing a nurse passing surgical instruments to the surgeon. She also perceived anesthesia and echocardiography machines located behind her head. Beauregard was able to verify that the descriptions she provided of the nurse and the machines were accurate. Furthermore, during the experience, the woman reported feelings of peace and joy, and seeing a bright light. Although this is an individual case, it does illustrate the possibility of conscious mental activity during circulatory arrest, which biologically is what happens when we die.

Such stories of people with accurate memories of what occurred during a prolonged period of unconsciousness after the process of death has started comprise a major part of the experiences reported around the world. Indeed, they raise the question: Are these experiences real? To many people, these stories have become fodder for religious and philosophical debate. For some scientists, they may be rejected as unusual hallucinations, but to many scientists who have studied them in detail, they provide a window to awareness and mental activity in the death state, paradoxically binding us all together and freeing us up to make a scientific exploration into the mystery of what happens to human consciousness after we die.

CHAPTER 7

The Elephant in the Dark

THE ELEPHANT IN THE Dark" is a famous story that probably originates from India with variants of it existing in many cultures. A beautiful variation of it was put forth by Rumi, a Persian poet in the thirteenth century. One day a group of people in India who have never seen an elephant hear that an elephant has been brought into their town. Excitedly, they rush to where the elephant is being kept to see this magnificent creature for the first time, but the only problem is that it is night and extremely dark. Undeterred, they wish to experience the elephant for the first time no matter what, but because it's dark and they can't see anything, they decide to feel the animal with their hands and determine what it is.

One person touches the trunk: "It's a water spout." The second touches its ear: "It's a huge fan." Another rubs the elephant's leg: "It's a pillar." And one brushes his hand across the back: "No, none of you are right, it's actually a throne." Because they had never been presented with this animal before and because they could not see the entire elephant, they drew conclusions based on the limited amount that they could perceive through touch. They couldn't understand the entirety of what they were feeling, because the elephant was much more than what each individual had perceived.

The same is true of anything new presented to people. Individuals see the world through the prism of their own eyes and draw conclusions about a subject based on what they know or believe to be true. "The Elephant in the Dark" provides insight into our limitations when it comes to determining the nature of the truth and the reality of a given subject. It also demonstrates how these limitations impact our ability to provide expert opinions where there is inadequate information, whether we are aware of those inadequacies or not. Furthermore, it highlights the need for tolerance and respect for different opinions. Interestingly, perhaps since humans had never thought it would be possible to go beyond the threshold of death and then come back and describe it, one can imagine that after people first described their unusual experiences, the reactions to the experiences were much like the reactions of those people trying to "see" the elephant in the dark in India. Throughout the more than thirty years since they were described, many have tried to "touch it" and have described what it is from their own perspective. Before these experiences had become well known, some people, due to cultural or personal experiences, had come to believe in the concept that some sort of existence continues after death, whereas others did not believe in such a possibility. More and more people began describing something that resembled what people had mentioned about the so-called afterlife. This was the case when the world was first made aware of the experiences of people who had died and been brought back to life.

Before Raymond Moody first made the general public—and hence the scientific community—aware that people who had actually died had experiences, people didn't know about the phenomena. Aside from the odd, anecdotal case, nobody had seen or heard about NDEs before, and by and large, stories of these experiences weren't public knowledge or part of the discourse. When these first reports of mystical-sounding experiences came out, because science had not studied them, people interpreted them based on their own personal visions of life and death and assigned meaning to them based on their own views. No matter what people ascribed to them, these experiences put us in an uncharted territory where we could not have gone before the advent of modern resuscitation science.

Whenever we look at something in life, our ability to "see" it in its entirety is not absolute. We can't fathom every permutation because our levels of perception are not adequate to comprehend everything presented to us. The fact is that our three-dimensional brain and the powers of perception that arise from the brain are very limited. We cannot even see everything that exists in a room. We can see electromagnetic waves coming off objects in the visible light range of the electromagnetic spectrum, but we cannot readily perceive waves coming off objects where they are beyond the visible light range and thus we assume they don't exist, although certain animals have the ability to see beyond what we can see. People also look at the same event and interpret it differently based on their personal abilities to perceive, which often reflect a limited segment of the entire reality of a given subject.

What happened with NDEs and the reactions that people had to them after they were described is very similar to Rumi's tale. For example, if one always believed that everything in existence, including human experience, can be reduced to and specifically explained by the physical processes known and understood by humans in any given time, then that is the person's vision of everything in life, including these experiences. This is the so-called reductionist view. There can be nothing beyond that. Therefore, the person will attempt to find ways of explaining something new presented to him or her (such as these experiences that take place during a period after death) based on the individual's personal beliefs and prior models of thinking. The problem here is that there are so many things that we have not yet discovered in any era, and until we discover them, things may sound magical and strange, even mythical. For example, it may be that what we discover at times is entirely new and cannot be reduced, based on what has been understood by the science of the time. In the analogy of the elephant, if one isn't willing to accept there could be an entirely new animal called an elephant but tries to simply reduce the elephant to an animal that has "a fan" and a "water spout" and a "pillar" and a "throne," one cannot get the entire truth. A person with such a mind-set might say that since oxygen levels must be reduced or carbon dioxide levels increased in the bloodstream when someone is close to death, it must be one of these factors that is causing the experience to take place, and the person would therefore conclude the experience must be a hallucination. Like the person who deemed the elephant a water spout and could not interpret it in its overall context, this person may be drawing a conclusion about the experience by interpreting only one small area and not taking in all the evidence.

The same is true of other snap interpretations of the experiences that people have after cardiac arrest. Some people attribute psychological theories to the experiences, such as a psychological reaction to a fear of death. This interpretation is limited because it does not account for what is happening in the brain of someone who has gone through cardiac arrest and died. In these cases, it clearly can't be an adequate explanation since people who suffer a cardiac arrest, like Joe Tiralosi, don't know when it is going to happen, so they haven't had time to imagine something pleasant psychologically to comfort them through death. When we consider the views of others who have called the experiences a hallucination brought about by a lack of oxygen and nothing more, we find many problems. Least of all, the problem is that people suffer from the effects of a lack of oxygen in emergency rooms every day and don't have experiences that resemble what people have recalled from a period of cardiac arrest. In fact, hypoxia (or lack of oxygen) is one of the commonest problems physicians face in intensive care units and emergency rooms all over the world. For example, many patients have pneumonia or asthma or other severe breathing disorders, all of which can cause oxygen levels to drop down to incredibly low levels, yet people don't experience anything that resembles what has been termed a near-death experience. In fact, lack of oxygen leads to delirium, confusion, and coma due to a reduction in oxygen delivery to the brain. However, people with NDEs have lucid, well-structured thought processes with reasoning and memory formation, and they are quite the opposite of delirious. In short, the entirety of the phenomenon cannot simply be explained by one theory. Everything must be considered and reviewed in full to produce a complete picture of the meaning of the experiences.

People who are critically ill clearly will potentially have many different types of experiences, but it does not make sense to clump these experiences all together and label them as the same thing. This creates a lot of discussion and debate—some of it healthy, to be sure—but not a comprehensive study of the situation. That's why, for example, I believe that for research purposes in cases where people have suffered a cardiac arrest and been resuscitated we need to change the term *near-death experience* to *actual-death experience,* because in the cardiac arrest phase, we can say exactly what is happening to the body's physiology. The idea that someone suffering from meningitis had the same experience as someone who had almost bled to death after a motor vehicle accident or someone with an actual cardiac arrest is not accurate and specific enough. For starters, the biology will be very different. There may be similarities, but they are not the same thing, just as to someone who has never seen an elephant, a rhinoceros, and a hippopotamus, the three share many similarities, but to those who have more knowledge, they are clearly different. This is why we need to clearly define what we are talking about.

REPORTS OF PEOPLE HAVING interesting, unusual experiences and mental recollections after coming close to death date back thousands of years, though in ancient times they were not called NDEs. Perhaps the oldest reference to a similar experience occurs in Plato's *Republic,* written in the fourth century B.C. Here, an ordinary soldier suffers a near-fatal injury on the battlefield and is revived in the funeral parlor. He describes a journey from darkness to light, accompanied by guides, a moment of judgment, feelings of peace and joy, and visions of extraordinary beauty and happiness.

Other historic examples recorded centuries later mirrored what Plato had written. One of the most compelling was in a work of art by Hieronymus Bosch, a famous fifteenth-century Dutch painter. In his painting entitled *Ascent to Empyrean,* Bosch painted what looked like a typical case of what Moody later described as an NDE: angels escorting the departed people down a tunnel toward a bright light. It wasn't clear what Bosch himself had experienced or whether someone had described it to him, but someone who wrote to me with his own experience in the late 1990s said, "This very much resembled what I saw."

The first published scientific study examining people who had experienced a close encounter with death was released in 1892 by Albert Heim, a distinguished Swiss geologist and mountaineer who during his lifetime received many scientific accolades for his geological research in the Alps. Heim himself had survived a near-fatal mountain-climbing accident. During the accident, he found himself calm and unafraid of death and came away with a renewed, positive outlook on life. Fascinated by his own experience, he collected thirty firsthand accounts from survivors of similar accidents and compared them, though he had not yet determined that they were what Moody would almost a century later term near-death experiences. Heim found that he and all the survivors had similar experiences. Facing death, they felt no grief, anxiety, pain, or despair, but rather a calmness that evoked clarity washed over them.

Before reports of mental and cognitive experiences from a period close to and around death became more widely publicized in the 1970s, almost no one had heard of the phenomenon of having such profound experiences when close to death even though there were numerous cases of people who had reported these unusual experiences following their close encounters with death. One of the most famous people to discuss his experience was the psychologist Carl Jung, who described an experience that he had during an accident in his 1961 book *Memories, Dreams, Reflections*. There have also been many more historical accounts, and although when examining these records it is impossible to determine exactly how close each person had been to death, by and large they seem to resemble the experiences recalled by Moody in his collection of cases, as well as people that others such as I had encountered and studies later corroborated.

In the 1980s, reports also started surfacing of people who had experienced so-called negative NDEs. These people described experiencing frightful vacuums, demons, zombielike creatures, tortures, and other unpleasant experiences. It wasn't clear whether these had really taken place when the people had been close to death, or whether they had been due to the symptoms of a severe illness, such as excess carbon dioxide in the blood, which can sometimes give rise to such experiences and have a negative impact on the person experiencing them. This was another example of how experiences that were not clearly defined may have been grouped together under the NDE headline. For example, not only was their being near death not clear but it does not make sense to clump the experiences of people with different drugs and illnesses all together and call them all the same thing. Each medical condition has a very specific biology, and each medication and combination of medications is very precise—and therefore may affect people's experiences in a different and specific manner.

From the late 1970s through the 1980s, some limited research was undertaken on what Moody had termed NDEs. Scientists honed in on characterizing the different features of what people experience when close to death or after a cardiac arrest when people had actually died and been brought back to life again, and on finding out more about the nature of the experience and how often it occurred. At this time, researchers didn't fully recognize the importance of separating and distinguishing between experiences that take place under different circumstances. Consequently, all the experiences that took place in the setting of a so-called near-death encounter—often as defined by a history given by the patients themselves and not always corroborated through medical records or interviews with the doctors involved in the case—were classified as the same. The best account of the prevalence of NDEs came from a Gallup survey conducted in the United States in 1982. This survey concluded that NDEs had occurred in approximately eight million people, or 4 percent of the population, which to many seemed to be an astonishingly high proportion. Unfortunately, further surveys weren't conducted to confirm these findings, and there were no comprehensive data from other countries.

Although it was obvious that further research was needed to fully understand how prevalent the NDE was in society, the 1982 Gallup survey indicated that it was far more common than most people had thought. Researchers set out to investigate the people who reported having NDEs. They broke them down into categories that influenced people's thoughts and feelings, such as culture, religion, personality, intelligence, and drug use. They then set out to determine if certain types of people were predisposed to having the experience and if different people experienced a similar phenomenon. The key would be in the connections—or lack of connections—between the groups.

IN ORDER TO DETERMINE if an NDE was a medical and scientific issue, studies were conducted to determine whether variations occurred in the experiences of people from different cultures and backgrounds. Were those who followed a certain religion more susceptible to seeing something they believed in after they died? Did having a preconception of the afterlife carry over into an NDE? And what about those who had no religious beliefs—were their experiences similar to those who believed in an afterlife?

Understanding how different cultures and religions affected NDEs, if at all, was critical in determining whether or not the experiences could be scientifically studied. If all the experiences reflected people's preconceived ideas, then this would perhaps support the notion that NDEs were based upon a person's background and did not constitute a universal phenomenon that transcended personal, cultural, and religious views. But if people from different walks of life were experiencing the same things, then the NDE was more likely a "universal" human experience shared by all irrespective of culture, creed, or religion, such as love for our children. Thus, this may be the universal experience most of us will likely have when we go through death by natural causes.

In the 1980s, researchers found that, historically, events closely resembling NDEs had been described by diverse cultures and differing religions from Bolivian, Argentinean, and North American Indian stories to Buddhist and Islamic texts. There had also been accounts of these experiences in China, Siberia, and Finland stretching over thousands of years. In more modern times, the experiences had also been described in other areas of the world, including India, South America, and the Middle East, though relatively little if any publicity had been given to the phenomenon. The most common features were consistent among all races, religions, locales, and eras: having an out-of-body experience; reuniting with dead loved ones; experiencing a vision of light accompanied by joy and peace; and witnessing a border or dividing line between the living and the dead.

In the cases recalled by people in non–Western cultures, it was found that although the central features were universally present, the interpretation of the experience seemed to reflect religious or cultural views of that culture. So while people from different parts of the world may have all felt peaceful and seen a tunnel, a bright light, or a being of light, they each ascribed a different meaning to the light—a meaning that was directly related to their cultural and religious backgrounds. In one study carried out in 1985, the experiences of sixteen Asian Indians were compared with those of Americans, and it was found that the Indians often encountered Yamraj, the Hindu king of the dead, while the Americans had not. Same experience, different interpretation.

From the accumulation of all the different studies, it became clear that the central features of what Moody termed NDE appeared to have been recorded throughout history, as well as across numerous cultures and religious groups, and that other phenomena such as deathbed visions also share many similarities with the NDE, particularly with respect to having powerful visions of seeing relatives who had passed away as if they were welcoming the individual. Those who held different religious faiths had the same experiences as atheists. The conclusion drawn was that though preconceived ideas including culture and whether or not one followed a particular religious belief may have influenced the interpretation of what the person saw during the experience, it was not a major factor in having an NDE or the nature of the experience itself.

Separate and apart from religion and culture, personality traits were also studied. In the 1980s, several researchers investigated whether a particular type of personality was more likely to have an NDE. One study focused on IQ. It compared extroverts and those with high IQs to those who were more neurotic and affected by anxiety. However, no significant differences were found. A different study in 1984 examined measures of attention, agitated tendencies, death anxiety, danger seeking, and psychotic personality. Again, the study found no significant differences between the two groups, suggesting that an NDE was not dependent on a certain type of personality.

Because adults have a broader base of knowledge and more world experiences, they would be more likely than others in society to describe what they had learned about what happens when we die. The children in the studies were often too young to have formed an opinion regarding what happens when we die, or even death itself. This group, however, was clearly the best test of whether the details of what people experience during the NDE was based on preconceived, learned ideas or whether it transcended cultural and religious views and personality traits and was indeed a universal human experience.

Research carried out with children in the 1980s demonstrated that many had, in fact, described NDEs, and their experiences shared many of the same features as those of adults—separating from the body, watching events, feeling peaceful, and seeing a bright light or beings of light. The difference was that these were described in a child's vocabulary, often during the course of play and sometimes over many months. Though their interpretations of what they had seen were based on their own levels of comprehension, it was nevertheless clear that they had similar experiences to those of adults. Even more significant was the fact that some of the children studied were just two or three years old when they had their experiences. This group was certainly too young to have had any concept of death or the afterlife, yet they described experiences similar to those of adults.

Based on these studies, researchers were drawing the conclusion that NDEs crossed cultural, religious, personality, and even age lines. Today, scientists by and large accept that NDEs do in fact occur. What has remained a point of contention is their significance and what they really mean.

As it became clear that NDEs were occurring in similar fashion across every conceivable boundary, a small number of scientists and researchers began searching for explanations and testing different theories for the root cause of the experiences, all doing their work in parallel. Most of the previous examination had been done to specific areas and not evaluated with the other parts of the experience.

One previous explanation was that NDEs were a hallucination in response to the changes that occur in the brain at the time of death. Though the NDEs might seem real to those who had experienced them, some researchers believed that physiological and chemical changes accompanying the process of death might cause hallucinations and therefore account for the unusual experience. These changes included a lack of brain oxygen, increased carbon dioxide, the release of endorphins (the body's own morphinelike substance), and a specific type of seizure known as temporal lobe epilepsy, which is like an electrical storm in the brain.

The changes in the brain are largely based on the fact that processing visual changes and other deep experiences takes place in certain recognized areas of the brain. Although in most cases these are stimulated normally and lead to normal experiences, in disease states that cause hallucinations their activation can also lead to visions that don't correspond with any reality—which is the definition of hallucination. Abnormal stimulation of these same areas with chemically active substances, such as drugs like LSD, can also lead to visions and experiences that do not correspond to objective external reality. Someone who thinks he or she is flying after taking LSD isn't really flying. By analogy, it was said someone who sees visions of a so-called afterlife isn't really seeing that either. However, that line of thinking has many limitations.

For starters, all human experiences, whether real or hallucinatory, are mediated by a certain number of limited areas of the brain that are involved with both sets of experiences. The activation of specific areas of the brain can't determine the reality of an experience. It's not as if activation of one area means something is a hallucination and another area means it's real. For instance, feeling intense love is mediated by the same areas regardless of whether someone has the illusion of experiencing love or really is in love. The same is true of someone who experiences seeing a light; whether he or she is actually looking at a light or imagining looking at a light, the same areas of the brain become involved. Therefore, the first limitation is that finding any brain-based activation in any area can't tell us whether the experience is real or not. Nevertheless, researchers who have identified some possible difference in brain activity in people with and without NDE have then made the leap that the experience must be a hallucination. At best, this is scientifically a very weak position, because the reality or otherwise of any experience cannot be determined by chemical changes in the brain.

To explain how changes in the brain at the time of death may cause hallucinations, Dr. Susan Blackmore, a well-known British psychologist and early NDE researcher, proposed what became known as the "dying brain" hypothesis. This theory held that a lack of brain oxygen, which can occur during the dying process, might cause uncontrolled activity in the brain areas responsible for vision. This activity could trigger the illusion of seeing a light and a tunnel.

The dying brain theory was based on how the cells in the brain work in relation to vision. In daily life, the vast majority of our vision comes from the middle part of our field of vision with relatively very little coming from the peripheral fields. This is why we use our central vision for activities that require concentration, such as reading. The brain and the eye have far more cells devoted to analyzing information from the central portion of vision (since this is what we need to have detailed and focused vision) than the peripheral portion of vision. So what Blackmore was saying is that when we die, maybe the effect of a relative lack of oxygen causes all the cells in the back of the brain that are involved with vision to become active at the same time. However, since many more cells are involved with processing central vision than peripheral vision, overall activity and firing of all the cells at the same time may create an illusion with much more light in the center of our vision and the light diminishing as we go from the center to the periphery. She was hypothesizing that since people all have so many more cells devoted to vision in the middle and far less as they go away from the middle, they will get this illusion of a tunnel with a light in the middle of it.

This theory, however, has many limitations. If the dying brain theory were correct, then as the oxygen levels dropped in any person's blood, that person would gradually develop the illusion of seeing a tunnel and/or a light. In practice, patients with low oxygen do not report seeing a light, a tunnel, or any of the typical features of an NDE. In every hospital in the world and in every intensive care unit, doctors admit and care for patients suffering with the effects of lack of oxygen or hypoxia, whether it be acute or chronic. This is called hypoxemic respiratory failure, and a critical care physician such as myself has probably taken care of hundreds or even maybe thousands of such patients. People with critically low oxygen levels do not describe lucid, well-structured thought processes with accurate memories. As mentioned before, they either have no experiences because they are in a coma, or if the hypoxia is less severe, they become acutely confused and agitated and start to thrash around. But they certainly don't describe anything resembling an NDE. It is probably safe to say that millions of people all over the world may suffer with this condition at some point every year and do not have such experiences. There is also a huge scientific literature available that examines the effects of hypoxia, whether in hospitals or where it occurs at very high altitudes in mountain climbers. Again, NDEs are not a feature. In fact, this experience has not been reported by scientific studies as a result of a lack of oxygen, even though thousands of studies have examined the effects of a lack of oxygen.

There are also many other reasons that this is unlikely to be the case. Some may argue that this is not because of a relative reduction in oxygen but occurs due to an absolute cessation of oxygen being delivered to the brain when the circulation fails as in cardiac arrest and death. This is called anoxia (hypoxia is a relative reduction in oxygen, whereas anoxia is the absolute cessation of oxygen). Under these circumstances, patients lose consciousness immediately when the heart stops, and they go into a deep coma. The brain circuits also stop working within seconds, which should not enable lucid mental activity, thought processes, and memory formation to take place. In recent years, a number of scientific studies have been carried out in patients who have suffered a cardiac arrest and survived. By definition, they had all suffered anoxia to the brain during the episode, since they all had circulatory failure when their hearts stopped. If these experiences were merely an effect of anoxia, then everyone (or at least the vast majority) should have had the experience, because they all had the same problem. Last, a lack of oxygen does not cause brain cells to become active. As we will discuss later, it causes them to stop working and eventually die. Aside from the fact that finding a chemical change in the brain doesn't tell us whether an experience is real or a hallucination, it is unlikely that a lack of oxygen is causing the experience.

Though seemingly plausible from a theoretical perspective, from a practical perspective the lack-of-oxygen theory could not account for NDEs. Different researchers with different areas of expertise have tried to propose different chemical changes to explain how these experiences occur. The most well known of those is oxygen; people also have proposed other chemical changes to explain how these experiences could be occurring as a hallucination or an illusion, but they all share similar limitations to the oxygen theory.

As the brain is dying, various chemical changes are known to take place—many chemicals go up while many others go down. For example, we know that our cells respond to a situation in which our blood pressure drops critically (when we have entered the dangerous period of medical shock) by releasing huge amounts of adrenaline into the circulation. This is a last-ditch effort by the body to maintain and elevate the blood pressure and prevent circulatory failure. We also know that after death has taken place, levels of calcium inside the brain cells increase enormously while their levels outside the brain cells decrease. But aside from the fact that no chemical change in any part of the brain can tell us anything about the reality or otherwise of human experience, whether it be a so-called near-death experience or otherwise, the other key problem with any theory that solely focuses on brain-based chemical changes is that it doesn't take into account the physiology of what happens to overall brain function when the heart stops. This is the "elephant" we are talking about, where the individual changes are the "fan" or the "pillar" or the "throne." With respect to those people who had objectively died and whose hearts had stopped, we know that there is no measurable brain activity due to enormous changes that take place in the brain (as outlined in chapter 3). The brain goes into a flatline state, and it remains that way, generally speaking, even in those who receive resuscitation and CPR because only relatively very low levels of blood can actually get into the brain even with resuscitation. This lack of brain function continues sometimes for many hours even after the heart has been restarted because the brain is swollen and the pressure inside the skull has become so high that it resists blood getting into it easily even when the heart is pumping.

The other main limitation with any chemical theory is the underlying assumption that the experience must be a hallucination. To explain this better, think of another common human experience, namely depression. We know that depression is associated with numerous chemical changes in the brain, but as doctors we wouldn't dream of telling a patient who is suffering with profound depression that he or she is just imagining it, or that it is simply an illusion or a hallucination. Can you imagine if a doctor would say, "Well, although your depression seems real to you and I don't doubt that it seems real, it is just an illusion, a trick of the mind"? Clearly a doctor doesn't have to have experienced profound depression to accept that it occurs and that it is real to someone who is experiencing it. In the same way, although I, like many other doctors, have not experienced an NDE, I can't simply dismiss the experience. It is real for those who have experienced it, just as depression is real for those who have experienced it. I believe many would agree that finding a chemical change doesn't imply the experience is not real, and it also can't tell us whether it is truly real, in the same way that finding changes in chemicals in the brain of someone experiencing depression doesn't tell us about the reality or otherwise of the subjective sensation of depression that the person is experiencing. We accept it as real because it is real to the person experiencing it. Every human experience is mediated by a chemical change in the brain, but identifying this change doesn't negate or prove the reality of the experience.

Because brain function is so complex, scientists investigating NDEs looked for further chemicals that could be involved in the dying brain theory—that is, the theory that a chemical change in some part or parts of the brain involved with human experiences, sensations, and feelings could be causing the experience to occur as a type of hallucination. Drugs administered at the time of death seemed like an obvious explanation, but an examination of the medical literature doesn't support this possibility. Studies showed that many NDEs took place without any medications even being administered or that people with and without the experiences had had the same medications.

Scientists also began to look for the location in the brain where the experience may be occurring. Some scientists hypothesized that certain areas within the brain might be mediating the experience. It is well known that certain drugs, including amphetamines, ketamine, and phencyclidine, can lead to quite complex hallucinations. These drugs work by attaching to certain receptors found on the brain, which then activates those receptors and mediates their effects. For example, ketamine (a powerful anesthetic) and phencyclidine (LSD) attach to receptors in the brain, and when these drugs activate the receptors, hallucinations can result. In essence, these drugs stimulate the same parts of the brain that are activated when we see something in the normal course of life. This is again due to the fact that if anything stimulates a part of the brain that is involved with the perception of a particular experience, we may have a partial vision of that experience, in the same way that someone may feel a sense of love for others when intoxicated with alcohol. However, this doesn't negate the love that he or she may feel for someone when not under the influence of alcohol. Furthermore, because alcohol is able to stimulate a sense of love in some people, it can't be extrapolated to then suggest that the love that someone feels for someone else is a hallucination, or that the love that people feel during an NDE is not real or is hallucinatory or illusory, simply because the feeling of love can at times be stimulated by alcohol in people. By analogy, experiencing a bright light, love, or a beautiful place during an NDE cannot scientifically be dismissed simply because other circumstances or even the administration of certain drugs can also induce feelings of love, light, or any other feeling. If that were the case and in view of the fact that drugs can induce any feeling or sensation that we may care to think of as an experience, we would have to define pretty much all human experience as being hallucinatory or illusory.

The bottom line is that no brain-based chemical change can define whether a sensation or feeling is real or not. The brain regions involved with any feeling or emotion may not distinguish how they have become active, just that something has activated them. To simplify things, for instance, if someone feels depressed due to either the loss of a loved one or as a side effect of a medical drug, a chemical change occurs in the brain that affects neurons involved in depression in the same regions of the brain. This is why we can now prescribe antidepressants to people. These drugs adjust the chemicals in the brain and adjust the levels such that it is less favorable to depression and more favorable to happiness. Again, identifying a chemical change doesn't tell us about the reality of an experience, and neither does something else that can stimulate the same emotions.

Dr. Karl Jansen, a New Zealand brain researcher with expertise in the effects of drugs on the brain, studied the effects of ketamine and suggested that NDEs might be occurring as a hallucination through activation of the same areas of the brain when people are critically ill and deprived of oxygen. Testing this theory was another matter. Its major limitation was the same as the oxygen theory. Not only would identifying a specific receptor or chemical not determine the reality or otherwise of the experience, the receptor being discussed (the NMDA receptor) is very widely found in the brain and is involved in many other experiences and activities, such as memory recall, without causing hallucinations. Therefore, it would not be sufficient to assume that simply by virtue of it being active, an experience is a hallucination or real.

Another impediment to testing this theory, as with all chemical-based theories, is that after death has taken place, the brain has shut down and these cells are not in their usual state but are in fact undergoing their own process of death. They are severely abnormal and not in a state to mediate thought processes, whereas when someone has taken drugs and hallucinates, the brain is functioning and the cells are not dying, which is why he or she can experience these visions. If you give LSD or ketamine or any other drug to someone in cardiac arrest when the person's brain doesn't work, the patient shouldn't be able to generate any mental activity whether hallucination or otherwise, since you need a functioning brain in order to generate any thought process. This is an example of a theoretical model, which while interesting at first glance doesn't seem to apply to what is really happening in real life.

The other problem with the theory was that the hallucinations described by people who used drugs were not really like the visions described by those who had NDEs. Of course, any human experience may share similarities with other experiences, but that doesn't mean they are the same. There are only a limited number of individual experiences, such as happiness, sadness, joy, elation, and depression, that humans can experience. Often we may have a more complex experience that involves a combination of these feelings together, and as a result, different experiences may share certain common features. For instance, someone could see a light or even feel elated and happy under a multitude of conditions. It would, however, be hard to argue that because we may feel a sense of joy, happiness, and even elation during one experience such as the birth of a child, passing an exam in school, or even under the influence of certain drugs (or alcohol), consequently the joy, happiness, and elation that we experienced during our lives were not real.

Carbon dioxide has also been proposed as a possible cause for these experiences in the dying brain theory. In other words, the NDEs may be brought about by a change in carbon dioxide levels. The central problem with this line of thought is that alterations in carbon dioxide in the blood are also very common—as common, in fact, as a lack of oxygen. Anyone who has emphysema, chronic obstructive pulmonary disorder, or any lung disorder often may have highly elevated levels of carbon dioxide, but, again, these people are not having NDEs. Although a multitude of studies have involved changes in carbon dioxide levels, they don't indicate that people with elevations in carbon dioxide also have NDEs. In addition, experts in pulmonary disorders see many people with these conditions who do not describe having NDEs.

Another brain-based theory put forward was that the NDE was a type of seizure. Some researchers posited that the complex visions that took place during NDEs might have occurred due to activity in the temporal lobe of the brain, the area that processes vision and hearing. If this were the case, NDEs could possibly be linked to a condition called temporal lobe epilepsy, because the activity in this area of the brain had been shown to lead to complex hallucinations. During an epileptic seizure, abnormal electrical activity occurs in a specific section of the brain that leads to changes throughout the body. For example, if there were increased electrical activity in the areas of the brain responsible for arm and leg movements, the result would be a jerking of the arms and legs, which is what people commonly associate with epileptic seizures. It followed that if there were an abnormal change in the electrical activity of the visual areas of the brain, then people might see flashes of light or other visual images. That was the reason it is possible to have hallucinations during certain types of epilepsy, including temporal lobe epilepsy.

The fact that some people with temporal lobe epilepsy described hallucinations that shared some features of an NDE led some researchers to propose that an NDE was also a consequence of abnormal function in this area of the brain. But again, there was simply not enough evidence to draw a definitive conclusion because every single feature of an NDE can occur in multiple situations—you can see a light in your dreams or experience extreme happiness listening to music. If an NDE occurred as a consequence of overactivity in the temporal lobe, say due to a lack of oxygen, we would also expect to see some of the other features of overactivity of the temporal area, such as déjà vu experiences. However, these features did not occur in people who had NDEs.

Despite the fact that these brain-based theories—the dying brain theory, the activation of NMDA receptors, and temporal lobe epilepsy—and other less substantial ones, such as the release of endorphins that occurs when the body is under duress, were developed and explored by scientists, they did not independently or even collectively account for the NDE phenomenon.

Scientists also explored a different line of inquiry: psychological possibilities. There was a chance that even though an NDE appeared to be a real event, it was actually a series of thoughts that were a response to the stress of coming face-to-face with death. Simply put, if people thought they were about to die, they would then imagine experiencing events the way they expected them to occur. These fear-death experiences would be the product of the person's social, cultural, and religious beliefs. So people who believed seeing a light in death would save them headed toward the light, and those who believed they would end up in heaven soothed themselves by conjuring up an image of their prescribed nirvana. The major limitation with this theory was that people who suffered a cardiac arrest like Joe Tiralosi and had an experience were already medically dead and unconscious. They couldn't have had time to imagine something pleasant in advance because they didn't even know when it was going to happen.

BY FAR THE MOST controversial and difficult explanation to analyze was the idea that what people claimed regarding their NDEs might actually be real. Certainly this was what the vast majority of those who had NDEs claimed. Perhaps those who had an out-of-body experience and claimed to see doctors and nurses working on them from above really had been out of body, even though this was not something that science could really explain. Maybe it was neither an illusion nor a hallucination.

Those experts who believed that NDEs were a real experience argued that because the experience has been described all over the world, then it was unlikely to be a hallucination. If it were, people of different cultures would be expected to have different experiences, as their memories and thus hallucinations would be dependent on their life learnings. Another argument against NDEs being purely psychological or hallucinatory experiences was that they had also been described by children too young to have any concept of death or an afterlife. Finally, people had anecdotally reported being able to see things while hovering above at the ceiling in places and at distances that they could not have known about. These were areas that would need further sophisticated research.

Recent studies have attempted to delve further in the NDE, but because the term is so ill defined and frequently misused, the studies often veer away from the medical and scientific aspects of what occurs during death. Near-death experiences are such a hot-button topic that any time a study is released related to their possible cause, it generates publicity. Most of these studies try to find an association between NDEs and some narrow function in the brain as a way to dismiss NDEs as being hallucinations. Scientifically they all share the same limitations as other theories that assume the identification of a particular chemical change in a part of the brain can determine whether an experience is real or not.

One such study published in the journal *Neurology* in 2006 examined the role of sleep patterns and NDEs and concluded they could be related. The study, led by Dr. Kevin Nelson of the University of Kentucky, was entitled "Does the Arousal System Contribute to the Near Death Experience?" Nelson used a sample of 55 people that he found in an Internet survey who claimed to have had NDEs and compared them to a control group of 55 people who worked in a hospital. What the study concluded was that the 55 who had NDEs had certain symptoms during the day that suggested they suffered from alterations in the parts of the brain involved in sleep—in simple terms, a sleep disorder.

Another study that received widespread attention in the field was entitled "Carbon Dioxide May Explain 'Near Death Experiences.'" Published in the journal *Critical Care* in 2010, this study looked at people who had suffered cardiac arrests and found that those who had NDEs had slightly higher carbon dioxide levels. There are several problems with the *Critical Care* study's conclusion. First and foremost, as with all the other suggestions, a chemical change identified in a group of people cannot define whether their experiences are real or not scientifically. Furthermore, increased carbon dioxide levels may actually be a marker of higher-quality resuscitation methods; therefore, people who were successfully resuscitated would perhaps be expected to have higher carbon dioxide levels. Second, just because this sample had increased carbon dioxide levels doesn't mean that was the cause of the NDE. Finally, as explained, increased carbon dioxide levels are one of the commonest things that occur in hospitals. If carbon dioxide poisoning is responsible, then how come all people who have increased carbon dioxide in the emergency room don't report NDEs?

People are pushing the boundaries to extremes to try to put a label on this, and often it may simply be that they are labeling and calling the condition something it really isn't. This comes down to the importance of having clear and strict definitions but unfortunately neither the term *near-death experience* nor the term *out-of-body experience* has a strict enough definition, which means people could take something that shares similarities and call it a near-death or out-of-body experience. This is one of the main reasons I have advocated that researchers should try to focus their research efforts on people who have had cardiac arrest and thus objectively have gone beyond the threshold of death. Under these circumstances the biological processes that accompany the brain and other organs are well understood and are the same. It is the pathophysiology of cardiac arrest. Without this focus, we risk losing sight of the entire picture (as we saw with claims that carbon dioxide was responsible for NDEs). To underscore this, in one study researchers took healthy people and put special goggles on their eyes that showed them an image generated by a camera that was positioned behind them. Those subjects who had volunteered for the study could therefore only see the picture shown by the camera that was fixed on their own back. The people were then left for hours looking at that image of their back. After a while they became so used to that image that it felt like they were looking at themselves from behind. The researchers had created an optical illusion that seemed real to those people. Then the researchers pretended to attack the camera with a hammer. The people were startled because for an instant they felt like someone was attacking them from behind. Although fascinating in terms of explaining how an optical illusion could be

re-created using a camera and how our eyes may acclimate themselves to seeing things differently, this experiment didn't seem to
indicate anything much regarding the out-of-body experiences that
researchers had studied. It was almost as if the researchers had taken
something completely different to an out-of-body experience and
then labeled it as an out-of-body experience. Nevertheless, the researchers concluded that they had reproduced an out-of-body experience in the laboratory—and it was just an illusion. Of course, this
is not even close to what someone who is critically ill, suffers a cardiac arrest, is resuscitated, and then describes hearing conversations
or seeing events in an out-of-body experience would go through.
Furthermore, no one actually had an out-of-body experience, or at
least not an out-of-body experience that those of us who have studied it have seen. In an out-of-body experience, the person describes
a perception of separating from the body and looking down and
then recounts exact details regarding events that were really happening. During a cardiac arrest, this is often happening when the
brain is highly disordered or perhaps not even functioning. None of
these subjects had a sensation of separating from themselves or actually could describe anything else that was happening in the room
other than what they were forced to see through the goggles. I am
sure the researchers were quite genuine in their attempts, but clearly
they were studying something entirely different. I even wondered
whether they had actually ever met and interviewed people with
out-of-body experiences. These were certainly not out-of-body experiences. However, it was widely reported by the global media that
out-of-body experiences had been re-created in a laboratory and
that they were probably hallucinations!

In another study that attempted to explain NDEs, researchers
took people who were in a hospice dying and connected a machine
called a BIS monitor, which measures electrical activity, to their
brains. These people were clearly not going to be resuscitated as
they all had terminal illnesses and were dying peacefully and with-

out medical intervention. What the researchers noticed is that in the last few minutes before these people died, there was a surge of electrical activity in the brain that was transient and then went down. As already discussed, after the heart stops, electrical activity in the brain stops due to a lack of blood flow. Therefore, this part of the study made sense. Interestingly though, the researchers concluded that these people were probably having an NDE and that's why they were getting this surge of electrical activity. Obviously, this conclusion doesn't make any sense because we have no idea if they actually had an NDE because the individuals were all allowed to die peacefully and therefore nobody talked to them to find out if they had indeed had any experiences at all. More important, there could have been hundreds of reasons for the surge in the electricity—for the record, the most likely is the influx in calcium we have discussed that happens to brain cells around the time that people die; this would cause a major spike in electrical activity since calcium is moving in and out of the cells. As a researcher, I found it difficult to see how it could be concluded from this study that the people had an NDE or any experience for that matter!

These days any study that is labeled NDE is sent around the media outlets and Internet and often picks up a life of its own because there is such a heightened interest in solving the mystery. But the studies are often very narrow and almost uniformly inconclusive.

SCIENCE HAS DETERMINED THAT NDEs and, more important, ADEs do occur, and society is slowly catching up to this fact. So why do so many people often simply dismiss them as hallucinations, illusions, or fantasies? Although the reasons are probably quite complex, they relate to the need to define what reality actually is.

People's realities are determined and defined by what they know, and this reality has conventional boundaries. Often we have been conditioned to define reality as that which we can see, touch, and perceive with our five senses. We've been told that we don't know whether another reality exists beyond that. Any personal experience, including an NDE, may be very real to the person who has experienced it, but since others have not had that experience, they cannot say whether or not it is real. However, if everyone in society shares a particular experience, such as love, then society gives meaning to that experience and concludes that it is real. This is in fact how reality is determined.

Reality is not neurologically determined, then, but rather it is largely socially determined. People assign meaning to different events, phenomena, and observations. Since meanings are arbitrary, people can also change them, and therefore when circumstances change, the older definitions can become outdated. This is typically what happens in science. As more information becomes available, our definition of reality changes. In these circumstances, even though older definitions may no longer apply, changes are often met with a great deal of resistance. The source of progressive ideas is often vitally important in determining whether or not they gain acceptance. If the person proposing the ideas is highly regarded in his or her field and has a large number of followers, then the individual's views will be more widely accepted than those of someone who is perceived as being less important.

Most people agree that there exists a world of "reality," and in this reality is a world of objects and phenomena that is external to and independent of us and that we interact with. This reality has existed before us and will continue to exist after us, regardless of whether we believe in it. The laws governing the universe are a perfect example. Consider that all the principal scientific discoveries we have made throughout the centuries have always existed, yet it has taken us thousands of years to discover them. Electromagnetic waves have always existed whether people believed in them or not. However, it has been only in relatively recent times that scientists have discovered their existence and put them to use in daily activities that we take for granted, such as watching TV, listening to the radio, or e-mailing.

If you think about it, there must be other dimensions of reality that we simply cannot perceive with our five senses. Electromagnetic waves are a perfect example. They existed one thousand years ago, one hundred thousand years ago, even a million years ago, but we only discovered them just over one hundred years ago. If someone had proposed to people two hundred years ago that an "invisible" wave existed that can carry sound and picture as clearly as we can see and hear them thousands of miles away and be viewed by others, most people would not have believed the person. The reality of electromagnetic waves and the potential to use them has nevertheless always existed. Thus, there is a type of reality that exists whether or not we believe in it. Logically, it would be acceptable to conclude that there may be many other realities that we are not aware of, but do exist.

Another important reason for our limitations is that not only are all our senses limited, but so is our brain. The brain and the five senses that are connected to it can be likened to a very complex computer that is able to perform complicated computational analysis of data received from its sensors. But the sensors that we use to detect things in the outside world have limitations, as does our computer that analyzes all the incoming information. The brain, like any computer, is limited in its ability to process information by the hardware and software it has. If an external reality lies beyond the abilities of our brain to detect and interpret it, we will not be able to comprehend it.

We have all experienced the limitations of our brain's ability to perceive and comprehend. A simple example is our ability to interpret what we see when we look at visual illusions. Sometimes the brain simply cannot process all the information coming to it from the senses. Therefore, from a scientific perspective and objectively speaking, we cannot simply assume that people's NDEs are illusions or hallucinations, even if this is what sits more comfortably with our current neuroscientific models than the alternative possibility.

It could also be that we need a new science or a paradigm shift to explain a relatively new discovery. To understand NDEs, we must look beyond what exists. The way to start is to examine in a more objective manner the experiences that people have reported, ground them in clearly defined science, and find ways to test them more objectively in the future.

SINCE 2000 SOME RESEARCHERS, including me, have started to focus more on the cognitive experiences that people have during a cardiac arrest and hence death. This is no longer ambiguous and vague. Although most still call these experiences NDEs, from my research and work, I have determined, as stated earlier, that they are actual-death experiences, or ADEs. This is an important distinction, especially for research because the definition of an ADE is not vague like an NDE. An ADE is taking place during a specific biological setup where someone's heart has stopped. In the last ten years, at least five independent studies have been carried out and published in the scientific literature that have confirmed that 10 to 20 percent of people who have had a cardiac arrest and been brought back to life have had some cognitive and mental experiences from the period after death has started. The nature of the experiences is actually similar to that of NDEs, but these experiences are probably better termed ADEs for reasons I have explained. What we have found is that while, generally speaking, the people who have these experiences may recall a number of different features at the other end of the spectrum, 80 to 90 percent of people who survive a cardiac arrest have no recollections. So the question becomes, Why doesn't everyone who has been resuscitated have these experiences?

Although no one knows for certain, more than likely the answer is that some people who are able to recall their experiences have more resistance to the effects of the lack of oxygen delivery to the brain as well as the inflammation that engulfs the brain during the postresuscitation period with respect to their memory circuits.

There may be some people who are able to better recall things than others, despite the effects of a lack of oxygen and brain injury that would ordinarily wipe out all memory circuits. This is why we have a spectrum of memory recollections from very extensive and detailed recollections where people may recall seven or eight complex features to recollections of people like Joe Tiralosi who only recalled two features. By analogy it may be similar to what happens when we try to recall our dreams. We all dream every night, but for some reason some people are more prone than others to being able to recall their dreams. It may have something to do with the memory circuits of the brain. In cardiac arrest, we would expect all memory circuits to be down, because the brain isn't functioning; thus, the fact that some people do have memory recall is itself a paradox and suggests that a person's mind and consciousness may be able to continue even after he or she has crossed over the threshold of cardiac arrest and hence death.

Science has determined that people have experiences after cardiac arrest and death, but even today there are still people who question the experiences that people who have been resuscitated have had after death. Despite the presence of growing scientific data, they continue to dismiss NDEs as hallucinations, illusions, or fantasies. It is important for people to understand that the reality of an experience cannot be determined by which chemicals in which parts of the brain are involved because essentially the same areas are involved in both hallucinations and real experiences. Reality is the meaning we give to our experiences as determined by society. The main reason is that there are still very strongly held, definitive beliefs on life and death, and therefore anything suggested about what happens when we die is often confined to personal views. Again, to understand these experiences, we must examine NDEs that people have reported so that we can ground them in science and find ways to test them in the future.

Finally, although the subject of death and what happens when we die has traditionally been perceived as a religious or philosophical question, it is clear that it is a field of knowledge that requires the objectivity of science. We should remain impartial and be willing to accept whatever our inquiries bring up without being fixed to what we have been conditioned to perceive as reality, since knowing the reality of anything is difficult and that is what science should be concerned with—not what is perceivable at a given time based on a given fixed framework that is defined as science. As we have seen, while concepts such as consciousness, soul, or the afterlife have traditionally been considered nonscientific, scientific progress—and in particular the quest to save lives and understand death and hence conquer it even after it has set in—has made us reevaluate some of our perceptions and try to understand what happens scientifically. Today it is becoming much more difficult to define or understand death without considering a person's consciousness or soul.

CHAPTER 8

Understanding the Self

Brain, Soul, and Consciousness

WHEN MICHELANGELO WAS MIDWAY through painting the ceiling of the Sistine Chapel in 1510, another Italian Renaissance artist, Raphael, was commissioned by Pope Julius II to paint four frescoes in the Papal Palace in the center of the Vatican depicting the branches of knowledge. Raphael's most famous work of the four, *The School of Athens,* anachronistically depicts a cadre of Greek philosophers who lived in different eras hundreds of years B.C. In the center of the painting, deep in conversation, are Plato and his pupil, Aristotle. While historians do not know what Raphael intended, the depiction of Plato and Aristotle at the center of philosophy in a palace of such religious iconography is telling, because we know that much of what has been debated for centuries regarding the mind, the psyche, the soul, and the essence of who we are has come down to the two differing views proposed by Plato and Aristotle.

In the painting, the elderly Plato, shown with a long gray beard, is carrying his book *Timaeus,* which discusses human beings' relation to the physical world and guided philosophical thinking for millennia. His vibrant student, Aristotle, is walking by his side, clutching a copy of his book *Nicomachean Ethics,* a treatise of how men's lives were indelibly influenced by the heavens. Plato appears to be pointing to the heavens, while Aristotle appears to be gesturing toward the earth. It is believed that these gestures represent the core of their philosophies, namely that Plato believed in a dual world of existence, while Aristotle's beliefs focused on the tangible. The two could have easily been discussing the subject of the human mind, psyche, or "soul," and possibly even the afterlife, because almost every idea that has been debated through the millennia up until today on those subjects was rooted in the discussions and debates that Plato and Aristotle had along with the works of many of the other philosophers in the School of Athens.

Although in today's modern society, the term *soul* has taken on a relatively imprecise and mostly religious connotation, in the time of Plato and Aristotle in ancient Greece, the soul—then called the psyche—had a much more precise and specific meaning. The subject of what the psyche was included everything to do with human beings—what made them be alive, what shaped their identity, and what led to their morality. The psyche was considered the distinguishing mark of all living things, and in humans, it included emotional states and mental and psychological functions, such as thought, perception, desire, planning, practical thinking, as well as moral qualities and virtues. Contrary to common perception, the early definitions of the psyche, the soul, or "the self" did not always include belief in an immortal psyche (or soul) or even an immaterial psyche (or soul). And though most people now associate the word *psyche* with the mind, while the soul is perceived as something vague, esoteric, and even religious, its true meaning is, in fact, the soul or "self." In studying the psyche, Greek philosophers were concerned with understanding what animates living beings, notably humans, and gives them their unique characteristics. Just as there are today, there were many different and conflicting theories about both the origin of the human psyche or soul and also what happens to it after death.

Plato's views have helped shape much of modern Western and Near Eastern civilization and philosophical thought. He believed that the material world that we live in, the physical world that we can touch, see, and experience with our five senses, though "real," is less real and less "perfect" than another domain of existence (that is imperceptible with our senses) and belongs to the realm of the psyche or soul. Thus, the soul was an immaterial substance. He proposed that corresponding to this material world was a parallel world of existence that relates only to the mind and psyche or soul. By definition, that domain is perfect and eternal. So for everything that exists in the world of "matter," he believed that there is a corresponding "perfect" blueprint or archetype that exists in another realm of reality, and that perfect reality corresponds to the psyche or soul of the entity whose material form is visible to us. In other words, even though the concept of a human being is perfect, the reality is that because our world is contingent and reliant on many other factors, what we see in this world is not always perfect and that real perfection exists only in that other domain. Humans can get diseased, for example, or an animal can lose its leg. Plato called this the theory of forms.

The scholar David Banach, a professor of philosophy at Saint Anselm College in New Hampshire, explains that Plato's theories sought to define the divide between perception and reality. "The world we perceive through the senses seems to be always changing," Banach wrote. "The world that we perceive through the mind, using our concepts, seems to be permanent and unchanging. Which is most real and why does it appear both ways? The general structure of the solution: Plato splits up existence into two realms: the material realm and the transcendent realm of forms."

Plato's position was that the forms of all beings are eternal and immutable, and it is these forms that "mold" the physical matter of the being. By analogy, Plato said that when you look at a horse in the physical world, the idea of that horse (a specific form) is perfect. There is a blueprint or a mold for what an ideal horse would be like. If something happens and the horse breaks its leg and therefore becomes defective, that occurs in the physical domain. But in that other domain, or dimension, there remains a world of perfectness—a world of perfect "archetypes" from which all physical manifestations of beings emanate.

Plato believed that what the Greeks called the psyche, which was later translated as the soul in English, is eternal. It is the ultimate substance from which—like the world—the typical body is created. Plato believed that the ultimate reality was that domain, and what we see in the physical domain is an imitation that isn't as perfect as the one in that other domain. Based upon that concept, everything that we are is a by-product of that world of reality, and our physical body and our brain are secondary components of that ultimate reality. The psyche or soul, in essence, is the more real matter than the brain. The debate among scholars and scientists over this central point—does the brain contain or create the psyche?—continues to flourish and provide new avenues of thinking to explore to this day.

To describe his theory of forms, Plato put forth an allegory about a group of people who are living in a cave. Imagine, hypothetically speaking, that these people have been chained to the walls of this cave since birth and cannot leave. Above them is a tall wall and at the top of the wall is a small gap where light shines through. Every once in a while people outside of the cave that the cave dwellers have never seen walk past and cast a shadow on the wall. As long as the cave dwellers are stuck in the cave, they will perceive these shadows as real, and therefore that perception of the shadows is the ultimate reality that they can fathom of those people.

Now imagine that one of the men breaks free and climbs up to see where the light is coming from. When he looks out, to his aston-ishment he sees people walking around who are not black and white but are three-dimensional and have different colors and shapes. Ex-cited at his discovery, he runs down to tell the cave dwellers that what they have been seeing is not the actual reality; although it is real, there is an even higher reality to it. However, because the other cave dwellers haven't seen it, they don't believe him. They think he has gone mad with this ludicrous idea. Their perception of what is real is separated from the true reality. Although for many people, just like the chained cave dwellers, reality is only limited to what they can perceive with their five senses, there is in fact a higher domain of reality.

Aristotle came up with a different line of thought. Many would describe Aristotle as being perhaps the first great biologist, because he is the one who first created the taxonomy of biology that has formed the basis of the classification systems we use to this day. Though he was Plato's student, Aristotle did not believe that Plato's view was correct. He believed that the matter and the form of a being cannot be separated but can be distinguished from each other, and that the form of any being simply results from the characteristics of its matter. Therefore, a horse may have a specific form, but that form arises from its physical matter and not from some other ar-chetype or nonphysical entity. He theorized that the psyche, or the soul, is actually a by-product of the activity of the physical matter.

By analogy, Aristotle said that the soul is to the body what vision is to the eye. If the eye works perfectly, then as a by-product you have the phenomenon of vision. However, vision isn't the same as the eye; it is the soul of the eye. Further, he said that the soul in humans—their thoughts, their feelings, their psychological makeup, everything that makes them who they are—is simply a by-product of the perfection of their physical matter. For as long as the body works, you have a soul. But when the body ceases to function, you no longer have a soul.

Clearly, these ideas are in stark contrast when applied to the question of the psyche, or the soul, and what happens when we die. Plato's view was that when you die, that's not the end because the body was not the principal matter of the soul and that the more real domain of reality is the psyche, or the soul, which will continue to exist after we die. Aristotle, on the other hand, believed that when you die, that's the end because the soul was produced by the body and therefore cannot exist apart from the body. However, despite the fact that Aristotle largely believed the soul does not survive without the body, he also believed that a part of the soul, the intellect, might if perfected be separate from the body and remain after death, while the rest of the soul perishes.

In addition to Plato and Aristotle, many other great philosophers like Pythagoras and Democritus also debated the nature of the soul and what happens to it after death. Though no consensus was ever reached, an enduring theory was put forward by Democritus, who was the first person to have been associated with the theory of atoms. Democritus's theory, called the atomist theory, was that everything is made of atoms and the atom is something that cannot be divided further (amazingly, this was proposed some twenty-four hundred years before modern scientists discovered the atom). According to him and his followers, our body, our psyche, and our soul are all made up of indivisible components, and when we die, those atoms disperse in a void. If our mind, psyche, or soul is also made up of so-called soul atoms, they disperse as well, and nothing remains. In some ways, this may be similar to the views of those people who now believe that when you die, there's nothing left; you are gone. In contrast, Pythagoras, who lived before Plato and whose theory Plato supported, also believed that the entity of psyche or soul, that which makes us who we are, is eternal; therefore, when you die, it is not extinguished.

Today, if you were to ask people if they believe in the existence of the psyche, they would say, yes, of course. They may even look at you in amazement for asking the question. Few people would debate the existence of the psyche, because the psyche is probably universally accepted to exist because it is generally believed to be synonymous with the mind. However, if you were to ask, do you believe in the soul?, you are likely to get an answer that corresponds more with a person's personal or religious conviction or lack thereof. Most people are unaware that the terms *soul* and *psyche* actually corresponded with the same thing, which is really the "self." The reasons for the current religious connotations and associations will become clearer, but in reality, if individuals really say they don't believe in the soul, it's like saying they don't believe in themselves because the word *soul* means nothing more than the self, which is who we are. The debate has been over whether the psyche or soul is a by-product of the body and therefore dies with the body, as Aristotle believed, or whether it is in and of itself a real entity that can be separated from the body at death and therefore continue to exist, which was Plato's view (and is called dualism because there are two dual realms of reality). So the question they asked is: Does that thinking, conscious entity who is currently reading and contemplating the ideas in this book, which we refer to as "me" in daily conversation, come about through a physical process in the body, or is it some other entity? If so, what happens to it when we die?

Today, many who believe in the idea of a soul that continues to exist after death in the form of an "afterlife" don't realize the origin of their beliefs can be traced to Greek philosophy and in particular to Plato. During the Roman Empire, Christianity emerged, and by the end of the Roman Empire, it became the dominant faith. This caused the Greek ideas about the soul from Plato and Aristotle and many others to merge with those of the Christian faith. The church largely supported Plato's views because they seemed to fit more closely with the Christian beliefs of eternal life and resurrection.

St. Augustine, the theologian whose writings helped shape Western Christianity, was influenced by Platonic philosophy. He considered the soul to be a "rider" on the body, suggesting it is the soul that is the true "self" and is immaterial. He believed that man was "a rational soul" making use of "a material body." Today, people have very firm ideas about what is life, what is death, what is the afterlife, and what is the soul, whether they are Christians, Jews, Muslims, Hindus, Buddhists, agnostics, or even atheists. So if we believe that when we die, our self becomes nonexistent and disappears, we may be following a tradition left by the atomists or Aristotle without being aware of it. Or if we believe our psyche, soul, or "self" lives on after death, we may be following the tradition of Pythagoras or Plato and the Neoplatonists. But the fact is that whatever personal view we have regarding the self, psyche, or soul, it is likely that belief has been debated and developed by numerous scholars dating even further back than the ancient Greek philosophers. Although some beliefs may not be what science eventually proves to be correct, others may be.

As STRANGE AS IT may seem, this idea of the nature of the psyche, or the soul, is being investigated by scientists. Today in science, it is referred to as "the problem of consciousness." In other words, how do our thoughts, emotions, feelings, and essentially everything that makes us into who we are arise? A few decades ago, this topic wasn't even considered to be in the scientific realm. Now, scientists can be divided into two broad categories: those who support a view that broadly speaking corresponds with Aristotle's view, and those who broadly speaking support Plato's view. The essential question comes down to, Does the brain create the mind, psyche, and soul, or is the mind, psyche, and soul—in other words that entity that makes us humans into who we are—separate from the brain but interacts with it?

Each camp has very distinguished scientists, though today there are more who support Aristotle's view than Plato's. The most notable Nobel Prize winner who supported the view that everything we consider the "self" or soul arises from the brain—that when you die the body stops functioning and therefore the soul is extinguished—was Francis Crick, who was the codiscoverer of DNA. The most prominent scientist in the other group—which considers the essential reality of the human psyche or soul to be a separate entity from the brain and body and believes that when you die the psyche or soul continues as a different type of matter much like an electromagnetic wave—was Nobel Prize–winning neuroscientist Sir John Eccles.

Today, science has largely come to grips with the mechanisms that lead to various signals and messages being transmitted in the brain and the connections between the brain and the rest of the body. But the question that still remains and fuels the scientific debate is, Where in the midst of all this electrical activity and chemical processes that we know take place in the brain do thoughts and the self lie? Since our true reality is that we are all thinking, conscious beings and it is our thoughts that lead to our daily actions, where do thoughts come from, and more precisely, how do they come about? How does the passage of electricity across a cell lead to feelings? When we experience a feeling, such as love or benevolence, or even jealousy, scientifically we can track all the pathways that mediate the feeling, but how do these chemical and electrical processes turn into feelings and thoughts?

David Chalmers, an Australian philosopher, has summarized it very well: "Consciousness poses the most baffling problems in the science of the mind. There is nothing that we know more intimately than conscious experience, but there is nothing that is harder to explain." In his books, Chalmers has called this the "hard problem" of consciousness. This is in contrast to the "easy problems," which essentially involve understanding the mechanisms that allow the brain to deal with the various sets of information that it receives.

Modern medicine has helped answer some of the questions regarding the relationship between thoughts and the brain; namely, the specific areas that are involved with certain feelings, emotions, and thoughts, but not the question of how thoughts are actually produced from brain cells.

Methods of analyzing thought processes now involve special brain scanners called functional MRI (magnetic resonance imaging) and PET (positron emission tomography) scanners. These work on the principle that brain cells have a constant need for blood, which carries with it all the vital nutritional substances that they need to work, including oxygen and glucose. So the scanners essentially detect and follow the movement of blood to various parts of the brain. This way, they can tell us which part of the brain is working more actively at any time.

In addition to detecting changes in blood flow, specialized scanners can also detect the areas of the brain that have increased their consumption of oxygen and glucose. So by following the changes in flow of blood and the consumption of oxygen and glucose to various parts of the brain, scientists can understand which areas of the brain are involved with certain thought processes. This is called "mapping" the brain. To do this, scientists place someone into a scanner and scan the brain while the person is having certain thoughts.

As I read what I have written, changes are constantly taking place in the flow of blood to certain parts of my brain. I am also enjoying music, and this pleasant feeling is accompanied by a reciprocal change in the pattern of blood flow to the part of my brain involved with this sensation. If I really get into my music and stop paying attention to the screen, the areas that had been receiving more blood while I was reading will now receive less blood, but other areas will then start to receive more blood. Interestingly, brain scans have shown that for any thought, many areas of the brain become active and that it is therefore multiple areas of the brain that mediate thought processes. This is a very important point. However, identifying blood flow changes or increased metabolism of certain parts of the brain during an experience doesn't answer the big problem, which is: How does a physical collection of cells give rise to conscious experience?

One method of studying this area that has been introduced by scientists in the last fifteen to twenty years has been the effort to determine the brain-based changes that take place under certain conditions and correlate them with conscious experience. In other words, scientists have attempted to study the biological processes that take place in the brain when someone has a conscious experience, such as seeing an object or thinking. These processes have been called the neural correlates of consciousness (NCC), which could thus also be called the neural correlates of the "soul."

These discoveries led some to believe they had discovered the "seat of the soul" in the brain and that clearly the soul was nothing much more than what they could see with their scanners. But an important point highlighted by other researchers and scientists was that just because something correlates with something, it doesn't necessarily mean that thing caused it. This is a basic law of studying correlations in any field of science. When any correlation is observed between two events, there are three possible explanations. Let's take the example of two actions, A and B. If there is a correlation, then either A is causing B, B is causing A, or some other process is causing both of them. So when correlations are found between brain-based events and conscious experiences, all the possibilities must be considered. Although we know that brain-based events correlate with thoughts, no one has been able to demonstrate whether A causes B, B causes A, or both. In other words, perhaps the brain-based events cause conscious experience, or maybe conscious experience causes the brain-based changes—or something else causes both of them.

In medical and scientific literature, a number of different theories have been proposed to account for how human consciousness or the psyche or soul comes to exist from the brain. The most commonly held view is that it is simply a by-product of brain activity, just as, for example, light arises from the action of electricity passing through a lightbulb or heat arises from the burning of coal. Hence they were not the same as the underlying processes that take place in the brain, in the same way that light coming off a lightbulb is not the same as the activities that are taking place in the lightbulb. As Aristotle would have said, vision is to the eye what soul is to the body. It is through the activities of the physical processes in the body that the soul arises.

Although experimental scientific evidence or even a plausible biological theory demonstrating how this can happen is still lacking, a number of different theories have been proposed to account for it. These theories have attempted to tackle different aspects of the problem of consciousness, such as how conscious experience may arise from brain cells, or how the different aspects of consciousness may bind together to form a single, unified sense. This is a particularly interesting and mind-boggling problem. We may not realize it or have ever thought about it, but all that we experience at any one moment is part of one conscious state, yet the different aspects of what makes us have one thought or feeling are actually mediated by many different areas of the brain.

Take vision, which is itself divided up into color, motion, and form processing. It is mediated by several areas in the brain at the same time. Now imagine that we are watching a movie and thinking about it at the same time, or experiencing a certain feeling of happiness or nostalgia. Each feeling or thought itself is also being processed by many different brain regions. Therefore, if we take a snapshot of what we are experiencing at any one moment, we realize that while each aspect is being mediated by a multitude of areas in the brain simultaneously, remarkably the "I" experiences it all as "one" and not hundreds of different and separate things. This is what neuroscientists call the binding problem.

Several views have been proposed to account for how conscious experience may arise and how the binding of consciousness into "one" may occur. It has been suggested that mental states simply arise from specific patterns of activity within networks of brain cell connections, or that this may be related to a specific pattern of synchronized and rhythmic electrical activity in networks of brain cells. Some others have proposed that consciousness emerges as a novel property simply out of the complexity that is going on among brain cells. Others still have further argued that conscious experience emerges when a critical level of complexity is reached in the networks of brain cells that are connected together. So although connections among a handful of brain cells may not give rise to conscious thoughts and feelings, conscious experience may arise when many more cells connect together, say a few million or billion cells. In addition to the mechanisms proposed in these theories, many other areas of the brain have been proposed as the potential neural correlates of consciousness. However, although very interesting, all these theories seem to share the same limitations.

IN GENERAL, SCIENTIFIC EVIDENCE to back up the concept that mind and consciousness, or in other words the psyche or soul, arise from the brain has come from the clinical observation that specific changes in function such as personality or memory are associated with specific areas of damage to the brain, such as those that occur after head injury or a stroke. This finding has been further supported by the results of studies using functional MRI and PET scanning, in which (as described above) specific areas of the brain have been shown to become active in response to a thought or feeling. However, although these studies provide evidence for the role of neuronal networks as an intermediary for the manifestation of thoughts, they do not necessarily imply that those cells also produce the thoughts. It still doesn't answer the question of whether Plato's concept of a psyche or soul that is separate from the body but interacts with it was correct or Aristotle's idea that it is the body that produces the psyche or soul. In fact, many scientists have argued that brain-based theories cannot fully explain the observed features of consciousness or soul.

The limitations of the conventional theories can be divided into four broad categories. The most obvious and most important limitation of such theories is that, first, they do not provide a plausible mechanism that may account for the development of consciousness, thoughts, and all that makes up the human psyche or soul from brain cell activity. The theories simply propose potential intermediary pathways that may be mediating consciousness and the soul but do not answer the fundamental question of how thoughts and consciousness or human experience may arise from the activity of neurons. This is a big challenge in neuroscience. We know we all exist as thinking, conscious beings and that if we take the original and more precise definition of the psyche and soul (rather than some of the vague concepts that some people have today of the soul), then we know it exists. But the question remains, How?

We also know that brain cells like any other cell can manufacture protein-based chemical substances. They can even be linked with electricity, but the nature and substance of thought (the amalgamation of which comprises the self) seem inherently different from electricity or a chemical or any protein-based substance we know. Most scientific theories simply say thought exists and it comes about from the brain but can't specify how, where, or why. This is a point that has been summarized very well by Professor Susan Greenfield. She concludes in her article "Mind, Brain and Consciousness": "Just how . . . the bump and grind of the neurons and the shrinking and expanding of assemblies actually translate into subjective experience . . . is, of course, another story completely." Second, another limitation relates to the issue of how brain activities that are distributed and are occurring within multiple different areas of the brain and more precisely billions of individual brain cells at any one time can eventually bind into a single unitary sense of self that leads to the notion of "I." Third, how do occurrences that are preconscious (in other words, chemical or electrical events that are continuously going on in the brain but are not part of our "conscious" awareness,

such as the effect of hormones, or other events that are instead occurring in the unconscious domain) become conscious, other than to say
that it somehow does occur at a critical point?

Finally and perhaps most important, we know that a fundamental part of our lives involves the notion of free will. We are judged
in society based upon our intentions and actions, and the brain-
based views expressed above cannot account for this. If correct,
they would mean that our lives would be completely determined by
our genes and environment and hence there would be no place for
personal accountability. Can you imagine the situation that would
arise if everyone claimed that everything they did was due to the
action of their genes in combination with their environment? No
one could be held accountable for anything.

THESE AND OTHER LIMITATIONS with the conventional views have
led some scientists to seek alternative explanations for consciousness. Stuart Hameroff, an anesthetist at the University of Arizona,
and Roger Penrose, a mathematician from the University of Cambridge, have raised many of the limitations of the theories above. In
particular, they argue that they cannot fully explain the observed
features of consciousness. They have put forward a theory using
quantum physics. Their theory is based on the principle that there
are two levels of explanation in physics: the familiar, classical level
used to describe large-scale objects, and the quantum level used to
describe very small events at the level of existence so small it is even
smaller than an atom (called the subatomic domain).

At the quantum level, superimposed states are possible—
meaning that two possibilities may exist for any event at the same
time—but at the classical level, either one or the other must exist.
So, for example, we can either go right or left, but not both. When
we make an observation, we are working at the classical level, so although there may be subatomic processes going on at any one time
with the potential of different superimposed states, when we make

an observation, the superimposed states must collapse into one.

Hameroff and Penrose propose that consciousness arises from tiny, tubelike structures made of proteins that exist in all the cells in the body, including brain cells, and act as a skeleton that allows cells to maintain their shape. They propose that these small structures are the site of quantum processes in the brain due to their structure and shape. They argue that consciousness is thus not a product of direct brain cell to brain cell activity, but rather the action of subatomic processes occurring in the brain. In support of their theory, they further argue that there are single-celled organisms, such as amoebas, that, despite lacking brain cells or any connected assemblies of specialized brain cells, seem to have consciousness and are able to swim, find food, learn, and multiply through their microtubules. Hence they suggest this may be a more advanced structure leading to consciousness.

So Hameroff and Penrose propose that consciousness, or what the Greeks called the psyche or soul, may arise from subatomic quantum processes occurring in the protein structures that make up the microtubules. Some have, however, argued against the theory by pointing out that microtubules exist in cells throughout the body and not just in the brain. Also, there are drugs that can damage the structure of microtubules but appear to have no effect on consciousness. More important, it has been argued that although the theory may potentially account for how the brain carries out complex mathematical problems, it still fails to answer the fundamental question of "how" thoughts, feelings, emotions, and what makes us into who we are arises.

This limitation of all the theories has led to the suggestion that human consciousness or the soul may in fact be an irreducible scientific entity in its own right, similar to many of the concepts in physics, such as mass and gravity, which are also irreducible entities. The investigation into consciousness has been proposed to be similar to the discovery of electromagnetic phenomena in the nineteenth century or quantum mechanics in the twentieth century, both of which were inexplicable in terms of previously known principles. Some, such as philosopher David Chalmers, have argued that this new irreducible scientific entity is a product of the brain, whereas others have argued that it is an entirely separate entity that is not produced by the brain.

The late Nobel Prize winner Sir John Eccles, considered by many to be one of the greatest neuroscientists in the world, was perhaps the most distinguished scientist who argued in favor of such a separation between mind, consciousness, and the brain. Eccles's theory was well described in his book *The Self and Its Brain*. He argued that the unity of conscious experience was provided by the mind and not by the neural machinery of the brain. His view was that the mind, consciousness, or soul itself played an active role in selecting and integrating brain cell activity and molded it into a unified whole. He considered it a mistake to think that the brain did everything and that conscious experiences were simply a reflection of brain activities, which he described as a common philosophical view. "If that were so," Eccles said, "our conscious selves would be no more than passive spectators of the performances carried out by the neuronal machinery of the brain. Our beliefs that we can really make decisions and that we have some control over our actions would be nothing but illusions."

Eccles further argued that there was "a combination of two things or entities: our brains on the one hand and our conscious selves on the other." He thought of the brain as an "instrument that provides the conscious self or person with the lines of communication from and to the external world" and that "does this by receiving information through the immense sensory system of the millions of nerve fibres that fire impulses to the brain, where it is processed into coded patterns of information that we read out from moment to moment in deriving all our experiences—our perceptions, thoughts, ideas, and memories."

According to Eccles, "We as experiencing persons do not slav-
ishly accept all that is provided for us by our instrument, the neuronal machine of our sensory system and the brain; we select from
all that is given according to interest and attention and we modify
the actions of the brain, through 'the self' for example, by initiating
some willed movement." However, he acknowledged that he was
still unable to explain how the mind carried out these activities and
how it interacted with a separate brain. This is a point on which he
was criticized by others.

Professor Bahram Elahi, a distinguished professor of surgery and
anatomy with a strong interest in the question of consciousness or
the soul and its nature, has expressed the view that although the
psyche or soul and the brain are separate, the entity we are referring
to as psyche or soul is not immaterial. Rather, it is composed of a
very subtle type of matter that, although still undiscovered, is similar
in concept to electromagnetic waves, which are capable of carrying
sound and pictures and are governed by precise laws, axioms, and
theorems. Therefore, in Professor Elahi's view, everything to do
with this entity should be regarded as a separate, undiscovered scientific discipline and studied in the same objective manner as other
scientific disciplines. He argues that as science is a systematic and
experimental method of obtaining knowledge of a given domain
of reality, then human "consciousness" or soul can and should also
be studied with the same objectivity. Each scientific discipline such
as chemistry, biology, and physics has its own laws, theorems, and
axioms, and in the same manner that which pertains to "consciousness" or the "soul" should also be studied in the context of its own
laws, theorems, and axioms. In his view, consciousness or soul is
also a scientific entity and a type of matter, though it is a substance
that is too subtle to be measured using the scientific tools available
today. Therefore, in his view, the brain is an instrument that relays
information to and from both the internal and external worlds, but
"consciousness" or the "soul" is a separate and subtle scientific entity

that interacts directly with it. Furthermore, as the human soul or consciousness is a separate yet entirely real entity that determines the true reality of a person, it continues to exist after death. Therefore, when one dies, one is left with the same level of knowledge, understanding, and perception as on earth. This is why people with actual-death or near-death experiences may interpret what they see based on their own level of education and thought. Thus, the level of perception of reality that exists after death is directly proportional to that acquired on earth. After a profound ADE, an atheist may maintain the same viewpoint, whereas someone of a particular faith may interpret what he sees based on what he already believed. Based on this school of thought, it is also therefore possible to educate oneself and hence correctly expand one's field of perception through the application of correct ethical principles during life on earth. The process by which one's depth of cognition pertaining to the realities that exist after death expands on earth (through the practice of correct ethical principles) is the purpose of our life.*

We know from the history of science that scientists have often been confronted with problems that had been unsolvable when examined using the scientific principles of the time. For example, when the Scottish scientist James Maxwell first discovered electromagnetic phenomena in the nineteenth century, electromagnetism had to be described as a scientific entity in its own right, as it could not be explained according to known scientific principles. It was many years later that the first radio waves (which are electromagnetic waves) were recorded by the German scientist Hertz, and now we have a whole area of science that is based upon them, not to mention numerous devices such as radio, television, microwaves, and infrared cameras. Maybe consciousness or that entity of the human psyche or soul that Plato, Aristotle, and numerous others

---

*For a wider discussion, please refer to The Path of Perfection *(Paraview, 2005) and* Spirituality Is a Science *(Cornwall Books, 1999) by Bahram Elahi.*

had debated is also not reducible in terms of currently understood mechanisms of brain cell activity, and its true nature will only be discovered when our science progresses.

AT THE SAME TIME that this debate on the nature of the soul has been taking place, from the days of the early Greek philosophers pictured in the School of Athens through the time of Nobel Prize winners like Crick and Eccles up until today, there has also been a related debate about what happens when we die, and by extension, what death is. As discussed previously, death has long been perceived as being irreversible and finite. But, as we are coming to learn and as has been described in some detail, in the twenty-first century we now have the means and technology to reverse death, and our ability is likely to expand even further in the coming years.

These new avenues of learning that have come about through advances in resuscitation science are very broad and have far-reaching implications that may not have been fully understood at the time they were first studied. Scientific exploration needed to be performed so that we could objectively study the entire process of resuscitation, its implications, and the questions it raised. One of the most profound questions was, If there is a period of time after death in which people could be brought back from death, what would happen to their minds, their psyches, their souls during that period that by definition corresponds to a time that is after death? Could we scientifically explore if the brain and the mind are separate—and therefore determine if the psyche still exists when the brain is not functioning? It would be time to set aside the philosophical and religious beliefs and see what the science would tell us. Science, by its nature, should have a completely open and objective mind-set. But how would we study this in any comprehensive way?

According to some experts, such as Professor Bruce Greyson, a psychiatrist from the University of Virginia, the situation parallels the developments in physics, which had to discover a new para-

digm of study when advancements were made. For centuries after
Sir Isaac Newton's discoveries, the entire world of physics relied
solely on Newton's laws of motion and the principles he laid down,
and the most eminent physicists of the nineteenth century believed
that there were only minor variations to be discovered. This was the
prevailing view until the twentieth century when the atom was discovered. Scientists began to notice that while Newton's laws were
applicable to the larger questions of physics such as the planets and
the stars, a problem emerged when they tried to examine the world
of the atom and a level smaller than the atom, a subatomic level.
Newton's laws did not apply to this subatomic world, and his equa-
tions and formulae didn't work. This caused physicists like Nobel
Prize winner Niels Bohr to search for alternatives. Their work led
to a scientific paradigm shift and the discovery of a whole new field
of physics called quantum physics, which relates to the world of
subatomic particles.

What this process showed was that with new discoveries there
was a need for a new science and a new definitional reality. The
Newtonian sciences worked for a particular area, but they weren't
sufficient to explain what was seen in the subatomic world. On the
macroscopic level, the old ideas worked, but when scientists looked
at those ideas in detail on the microscopic level that corresponds to
things that exist and are smaller than atoms, they didn't work. Some
have suggested that the same thing, as we will come to see, may be
true in the ideas that we have long held about the mind and brain
and consciousness.

The paradigms we hold in science may not always be absolutely
correct; they are the best that we can consider for any given time
with the information we have. When we make further discoveries,
we need new ways to explore these discoveries and amend our theories and specific paradigms. So although many (myself included)
previously assumed that death was the end, we have come to realize that death is not the end we once considered it to be. While all

cells, including the brain cells that mediate our thoughts, feelings, memories, and emotions, are going through their own processes of death, they are still viable for a period of time after we die. So now the question is, When someone dies, what really happens to his or her psyche or soul? Is it annihilated like many believe, or does it continue to exist for a period of time, and if so, for how long?

The discovery of resuscitation science would lead to the first reports of people who had died and been brought back to life telling us what they experienced. People related their astonishing experiences in the time after they had died and before they had been brought back to life. So although we once believed that there was nothing else to question about death, that it was black and white, like a century of physicists who said there was nothing to question about Newtonian physics, some have started to argue that we may now have the need for a paradigm shift regarding what happens when we die and the period after death. These realizations have only really started to come about after humankind's gradual discovery of the means to successfully reverse death itself—something that had always seemed impossible until now.

CHAPTER 9

The Afterlife We Know

FIREFIGHTER DON HERBERT WAS battling a raging fire inside a house in Buffalo, New York, on December 29, 1995, when the roof collapsed and trapped him under a pile of smoldering debris. Minutes before he nearly burned to death, members of his ladder team rescued him by pulling him out of a window, but in many ways it was too late—Don Herbert was already undergoing anoxic brain injury from having been stuck in a smoke-filled room for too long. Biologically speaking, essentially what was happening to his brain was the exact same thing that happens after a cardiac arrest and when people die: a global stroke—a stroke that is causing progressive cell damage in the whole brain and not just a part of the brain. Although potentially reversible at first, the damage eventually becomes irreversible. Herbert was rushed to the hospital, but by the time he arrived, he had slipped into a coma. He wasn't able to respond to any verbal commands, likely due to swelling of his brain. Although today medical interventions such as cooling may have prevented the progress from reversible to irreversible brain damage and cell death, not enough was known about it at that time, and he eventually progressed into a state of irreversible brain damage.

Though Herbert did regain consciousness for a period of time, he soon slipped back into a permanent disabled condition known as a minimally conscious state, confirming that after anoxic brain injury, a process of cell death and damage occurs over a number of hours and even days that eventually leads to irreversible brain damage and death unless an appropriate and timely intervention is made. *Minimally conscious state* and *persistent vegetative state* are two terms used to describe different points along the same spectrum of what happens to people after extensive damage to the brain areas involved with the process of consciousness and awareness. Traditionally, it has been taught that unlike a person in a vegetative state who does not exhibit any signs of consciousness, a person in a minimally conscious state, although also largely unresponsive, can muster some very limited responses to some external stimuli but is otherwise unable to perform any of the basic daily living tasks such as talking, walking, or eating without assistance from others. In short, the paradigm has always been that a minimally conscious state is a condition of severely altered consciousness in which minimal behavioral evidence of oneself or the environment is demonstrated that is completely absent in people in a vegetative state. Herbert's situation was so severe that he could not even be fed by others so he was given a feeding tube to keep him alive. After a period of time passed with him in this condition, a neurologist determined that he would not recover, and he was placed in a twenty-four-hour nursing home.

Herbert's case and many others like it demonstrate the devastating consequences of what happens to patients following anoxic brain injury whether from cardiac arrest and death or otherwise, without appropriate and timely interventions. His case also highlights the contrast between what happens to the Joe Tiralosis of this world and the Don Herberts of this world. Sadly, for every Joe Tiralosi, there are many more Don Herberts. The contrast is remarkable—

one hardworking father goes back home and back to work functional and contributing to society, caring for his family for many years after anoxic brain injury; another is left in a dreadful disabled state, unable to communicate, and unaware, while being cared for by society through taxpayer money. Aside from the ethical and moral cost to society and the devastating consequences to families, the financial burden to society is also not insignificant. It has been estimated that the cost of caring for each person suffering with long-term consequences of brain damage ranges anywhere from $600,000 to $1,875,000, and although the provision of health care is costly, clearly every brain saved will lead to significant cost savings.

Although it highlights the absolute need to establish a system that provides optimized, high-standard care for everyone so that there will be many more Joe Tiralosis and far fewer Don Herberts, this case and many others also shed light on one of the most mysterious, fascinating, and important questions that has intrigued scholars, researchers, and the lay public alike for years—the fundamental question of who and what we are and what happens to us after death.

Most would agree that while we all have an external physical appearance that changes with time, each one of us is fundamentally a thinking, conscious being with a unified mental life of his or her own. It is this that in essence distinguishes who we are above and beyond our external physical appearance. Our mental life, our "self," which remains "one" throughout our lives, whether young or old, contains an amalgamation of all our thoughts, feelings, instincts, memories, personalities, and more—in essence it contains everything that makes us into who we are. It is this entity—often referred to by scientists today as "consciousness"—that defines us. In Don Herbert's case, his consciousness or soul was everything that made him into who he was—it was that entity which was present and manifested as his vibrant self before his accident, but was largely absent after his anoxic brain injury. While in contrast to his body, which remained physically recognizable as Herbert, his consciousness, his soul, was no longer recognizable to anyone, and sadly it seemed to have been lost forever.

Ten years passed. Herbert's four sons grew up, and his wife continued to visit him daily and hold out hope of him one day talking again and being himself again. But as time passed, there was no measureable progress in his condition. He spent his days slumped over in a wheelchair, seemingly unaware of where he was or even who he was.

Then one day quite unexpectedly, Herbert regained awareness of who he was. He was more alert and though only capable of struggling to mention largely unrecognizable words, he managed to ask the nurses how long he had been gone and where his family was. When his wife and sons arrived, excitedly and in his own limited way, he kept asking how long he had been away. He was blind from the accident, but he recognized their voices. They told him that he had been gone for ten years. Utterly devastated, he began to cry in disbelief. But how amazing! Don Herbert's consciousness, his soul, had come back to view after almost ten years of absence. Although not fully recovered and still unable to communicate more than a few words at a time and even that with great difficulty, Herbert was nevertheless once again able to more meaningfully interact with his loved ones, and, more important, he was aware and conscious again, albeit severely disabled.

Sadly, Herbert eventually died after catching pneumonia. Though his story did not have a happy ending, his case and many others like it have raised profound questions regarding our understanding of consciousness, the concept of the soul, and what happens to us after death.

As already discussed, today much of our perception regarding what makes up who we are is shaped by what we have been conditioned to believe through our given cultures, our societies, or our own personal beliefs. But clearly, with so many different opinions and so many different societies and social groups, not everyone's beliefs can be correct. In trying to identify the correct opinion, we cannot even rely on the majority opinion, since history has shown that the correctness or falsity of a belief is not necessarily related to whether it is held by a majority in a given social group or society. This fact even applies to scientists and scientific beliefs. For instance, studies have shown that strongly held opinions, accepted by the majority in medical science at any given time, usually completely change as frequently as every twenty or so years. So at least with respect to defining the true reality of a given subject, one cannot necessarily always rely on the majority scientific opinion of a given time. Furthermore, when discussing a subject, much of what is being discussed is often lost in the imprecision of the actual terms used, which may have different meanings to different people and thus engender strong emotional responses that limit the ability to have a rational and reasoned discourse. This is particularly the case with understanding the nature of the self and what happens when we die.

Here is what we do know today. Through scientific study, we understand how brain cells produce proteins and different chemical molecules. We even understand how the profile of brain cell chemicals called neurotransmitters (chemicals that help transmit electrical signals between cells by moving in and out of cells) change with the changing of our thoughts and feelings. We also understand that proteins, chemicals, and electricity are fundamentally very different in nature to a thought or a feeling. Thoughts and everything else that make up a person's consciousness cannot be broken down into a protein or any other chemical molecule or even electricity. In science, we have not been able to come up with a plausible biological mechanism to account for how a cell or groups of cells working together (i.e., the brain) could possibly generate a thought or a collection of thoughts and hence ultimately the entity we call human consciousness. We know how cells make electricity through the movement of chemical molecules, but we just don't know how they could also make thoughts from electricity or chemicals. Much of what has been stated in support of the idea of the brain producing thoughts ignores this very important point, which, according to the Australian philosopher David Chalmers, is the "hard problem of consciousness."

What we do know is that the "self," consciousness, psyche, or soul—which includes the mind—does exist and is clearly linked with very specific regions in the brain that are wired and linked together through electrical impulses generated by the movement of certain chemicals in and out of cells. These closely linked regions in the brain, while not necessarily producing thoughts, demonstrate certain specific changes in association with our thoughts, feelings, and changes in our level of awareness of the external world. So if you feel sad or happy, jealous or benevolent; whether you enjoy listening to music or watching an opera or a game of football; and even when you feel love for your children, we can identify the biological correlates or "fingerprints" of all those feelings and thoughts through changes in blood flow and metabolism in the groups of cells that reside in specific areas of the brain. Of course, in the same way that a "fingerprint" is not a finger, the change in metabolism in cells or the change in blood flow in a brain region in association with a thought or feeling is not the same as a thought or feeling itself.

RESEARCHERS AND DOCTORS HAVE now delved into what happens to human consciousness and its relationship with activity in the specific regions of the brain. Putting aside the debate regarding the nature of human consciousness (or soul) and how it comes to be, and while acknowledging that brain cells may not have the specificity to be able to produce thoughts, there is much evidence that demonstrates that consciousness is nonetheless modulated by specific regions of the brain. Everyone has experienced this in daily life, particularly during sleep, which is the classic state where modulation of our consciousness takes place every day. When we sleep, we all lose consciousness and awareness of our surroundings, albeit temporarily.

A report by the National Institute of Health has summarized what happens during sleep very well. It states: "Nerve-signaling chemicals called *neurotransmitters* control whether we are asleep or awake by acting on different groups of nerve cells, or neurons, in the areas of the brain. Neurons in the brainstem, which connects the brain with the spinal cord, produce neurotransmitters such as serotonin and norepinephrine that keep some parts of the brain active while we are awake. Other neurons at the base of the brain begin signaling when we fall asleep. These neurons appear to 'switch off' the signals that keep us awake. Research also suggests that a chemical called adenosine builds up in our blood while we are awake and causes drowsiness. This chemical gradually breaks down while we sleep."

Many different neurotransmitters or chemicals are involved in the modulation of our level of consciousness and awareness in the brain, but interestingly they are not unique to consciousness or even the brain for that matter. They are some of the same common signaling chemicals found in other organs of the body. One example of a transmitter involved with our sleep/wake cycle is histamine, which is also involved in allergies. Histamine helps us remain conscious, awake, and alert, and this is why when we take some antihistamines, say for an allergy, we become sedated and sleepy. A reduction in histamine levels in the brain causes us to fall asleep and lose consciousness and awareness of our surroundings, whereas a reduction in histamine levels in the nose causes us to stop sneezing and reduces nasal symptoms during the allergy season.

Another transmitter that is involved with manifesting consciousness is acetylcholine, which is also found all over the body; when altered outside the brain, it can cause a dry mouth and diarrhea, among other things. Serotonin is yet another example of a transmitter that modulates our level of consciousness by making us

sleep, wake up, or become completely unconscious (depending on
the levels), but it also has many other effects too. Changes in the
level of serotonin in certain parts of the brain can cause depression,
while in the gut serotonin helps our intestines move more regularly.
This is why we give people with certain gastrointestinal disorders
such as irritable bowel syndrome "antidepressant" drugs because
they increase serotonin. But the same exact drug is also prescribed
to people with depression.

Finally (although I have not provided an exhaustive list), another modulator of our level of consciousness is dopamine; if altered
in the parts of the brain that deal with consciousness, dopamine can
impact our level of wakefulness and consciousness. However, if reduced in other parts of the brain (areas involved with movement), it
leads to Parkinson's disease. Outside the brain, dopamine increases
blood pressure and heart rate (among other things). In fact, in the
intensive care unit we commonly use dopamine drips to increase
the blood pressure in critically ill patients, while neurologists give
drugs that increase dopamine levels in specific parts of the brain to
treat people with movement disorders such as Parkinson's disease.

So what can we conclude? First, aside from the fact that a chemical is not the same as a thought, or a feeling, or the sense of awareness
and consciousness that we experience every day, we also can't say that
any state of consciousness or even feeling such as depression or happiness *is* the same as the specific chemical transmitter that modulates
it since those transmitters are found all over the body. However, we
do know that being aware, alert, and conscious as opposed to being
in a sleep, or an unconscious state and coma, takes place in response
to specific changes in neurotransmitters in very specific regions of the
brain that modulate our level of consciousness. Some changes cause
consciousness to "disappear" by leading us to go to sleep or lose consciousness entirely (this is what a coma is), while an alternative change
in the same transmitters causes consciousness to "reappear" by leading
us to wake up or become fully conscious again after a coma. Thus,

external awareness and a state of consciousness is orchestrated through activity in specific, complex brain circuits. This takes place through chemical regulation in the brain.

Researchers have discovered and charted the circuitry in the brain that modulates our consciousness. To simplify, it starts on the brain stem at the base of the brain, proceeds up the middle of the brain, and then goes into the frontal areas of the brain. The continuous circuit that activates it makes a person awake and alert, and if it shuts down for any reason, consciousness is lost, until the process causing the circuit to shut down is resolved. So consciousness can either be temporarily or permanently lost depending on what event has impacted or damaged the brain circuit that modulates consciousness.

Many events affect the brain regions that are involved in modulating consciousness, leading to a loss of consciousness and coma that, when reversed, lead a person to come out of unconsciousness. This is how sedative and general anesthetic drugs work. There are times when doctors want to purposefully shut down those pathways in order to medically treat a patient, such as during surgery. These drugs impact the pathways of the brain and take away pain perception, as well as take down memory circuits and consciousness circuits. That's why people can have surgery and not suffer horrible pain. Even though we can't see the patient's consciousness during that time, just as happens with sleep we know it is still there, and after the surgery we can bring the patient out from under the anesthesia and return the person to a fully conscious state. A commonly used drug that modulates the areas of the brain involved in consciousness is propofol. Although this drug received a great deal of media attention after the death of pop star Michael Jackson, critical care doctors and anesthesiologists have been using propofol in the ICU or before surgery for a very long time.

Aside from drugs, any medical condition that affects the chemical balance in those same specific brain regions will make a person's consciousness disappear too. In fact, any condition that leads to inflammation in the body and the release of specific inflammatory molecules in the body will alter the region of the brain that modulates consciousness. A simple everyday example of this is catching a cold or flu. This is why we become more sleepy and drowsy in general when we are ill. This is also why people may go into a coma when they become seriously ill and the level of inflammation becomes very high. These conditions include traumatic brain injury, vascular disease and stroke, brain infections such as meningitis, and liver failure. Infections cause a loss of consciousness and coma when viruses or bacteria (or the toxins produced by them) enter the brain and bind onto those areas of the brain that deal with consciousness. Severe liver failure causes coma due to an accumulation of toxins in the bloodstream (which the liver normally would dispose of, but cannot because it is not functioning properly). When these toxins reach the brain, they cause loss of consciousness by affecting the circuit that modulates consciousness. All these drugs and medical conditions share a common pathway: they cause changes in the chemical balance in those areas of the brain affecting consciousness.

Aside from the modulation of the chemical neurotransmitters that work on the specific regions of the brain involved with consciousness, whether due to a drug or due to an illness or even a loss of oxygen to the brain cells (anoxic brain injury), another mechanism leading to loss of consciousness is simply physical damage to the underlying cells that make up the brain circuit that modulates consciousness, such as occurs as a result of severe trauma. Thus, in Don Herbert's case, his almost complete absence of consciousness for ten years was initially due to the impact of a lack of oxygen being delivered to the cells, followed by the permanent damage that arose from anoxic brain injury in the specific areas of the brain that modulate consciousness. Even if the changes in the chemical profile are extensive, in many of these conditions the underlying brain cell structure is preserved and that is why when the condition (e.g., severe infection) has been resolved, a person's consciousness re-turns. However, cells permanently and irreversibly damaged cannot even respond to changes in their corresponding neurotransmitters, or they simply lose the ability to produce the neurotransmitters. It is as if the biological switch has been turned off permanently and thus human consciousness or the soul appears lost forever. As a result, Herbert had become permanently confined to a state of unconsciousness—his consciousness had "disappeared" and could never "reappear" because the circuitry needed to enable it to appear again had been damaged by the consequences of a lack of oxygen and was not functioning properly; when irreversible brain injury sets in, generally not much can be done. The key is to intervene before the damage has become irreversible. This is why Herbert's neurologist had thought his condition would be permanent.

However, incredibly and quite unexpectedly for Herbert and his doctors, after ten years, something changed in the structure of his brain that modulates consciousness, and his consciousness "reap-peared" and he became fully aware of his surroundings. It is as if the switch was turned back on again. It was not clear why and how his brain had partially repaired itself and enabled his consciousness to "reappear," but clearly during the previous ten years, Herbert's con-sciousness, soul, or psyche, the entity that made him who he was, his unified self, had not completely disappeared—it was always there, just not visible to the outside world and unable to communicate with it.

So while people in a minimally conscious or vegetative state have had permanent damage to those particular pathways involved in the modulation of consciousness, what is important to acknowledge is that just as when we are asleep, or after we have been given a general anesthetic, taken a heavy sedative, or suffered a severe infection, we can't say we have "lost" our consciousness in the true sense of the word (i.e., our consciousness hasn't actually left us and disappeared into the ether or even become annihilated). Instead, the circuit that modulates it is not active and so it is not visible and in contact with the outside world. Under these conditions, it also cannot modulate internal or external stimuli such as pain or experience memories, which is precisely why we don't experience anything, whether it be pain, memories, hearing, vision, or touch, when we are in a deep coma. It is almost like a person's consciousness—the self or soul—goes into a sort of hibernation mode. It exists but doesn't have meaningful interactions anymore. Therefore, in Don Herbert's case, while he seemed destined to stay in that state forever, something un-expected and incredible happened; somehow the damaged areas of his brain that modulate consciousness changed after ten years, and as that happened, his consciousness—his real self—"reappeared" and started to communicate with the external world, while having no idea how long he had been gone for.

For decades, in fact for as long as we have known scientifically, it was believed that once someone had completely lost his or her consciousness and the switch was turned off after irreversible brain damage, such as that which occurs after anoxic brain injury, there was no consciousness present at all. Thus, the dogma had been that people who end up in a persistent vegetative state after brain damage have permanently and completely lost their consciousness (self or soul). But a series of incredible and very interesting observations and discoveries a few years ago started to challenge this long-held belief.

WHAT IF DEEP INSIDE the brain, consciousness continues to exist even if there is no outward appearance of it?* What if conscious-

---

*Although there is a condition referred to as locked-in syndrome, where the problem is not due to damage affecting the regions that modulate consciousness, we are not referring to this condition. In locked-in syndrome, brain damage has caused paralysis while spar-ing the areas that modulate consciousness, so the person is not unconscious at all but just unable to move. Therefore the person appears to the outside world to be unconscious, while fully alert but trapped and unable to communicate. Here we are referring to condi-tions such as persistent vegetative state in which the centers that modulate consciousness have been permanently damaged and so consciousness is really absent.

ness is not lost at all, even in people with extensive brain damage who appear unresponsive and in a permanent coma? Despite the brain areas modulating consciousness seemingly being shut down forever, what if there are ways to awaken those areas and bring out the entity of consciousness? Scientists have started to explore the very real possibility that consciousness can go into sleep mode like a laptop computer and then can be awakened days, months, or years later in people who appear to have suffered damage to the areas that modulate consciousness. This research could result in major breakthroughs and have implications not just for neuroscience and resuscitation science but also for all of society.

Dr. Nicholas Schiff has done extensive work in this area. Schiff, who was named one of the hundred most influential people in the world by *Time* magazine, serves as director of the Laboratory of Cognitive Neuromodulation at the Weill Cornell Medical Center in New York. For a *60 Minutes* television report on the treatment of minimally conscious patients, Schiff studied Herbert's case and said that his awakening may have been caused by a Parkinson's drug. In other words, Herbert's damaged brain had become depleted of dopamine (one of the neurotransmitters that, as discussed, is involved in modulation of consciousness). Furthermore, Herbert was destined to permanently remain in a minimally conscious state due to the unrecognized depletion of the dopamine levels in the circuit that modulates consciousness—until he was administered a Parkinson's disease drug, which increases the level of dopamine in the brain. This had inadvertently corrected the deficit in Herbert's brain and hence Herbert's consciousness; his soul had suddenly "reappeared." His doctors and everyone else had assumed it was a sudden miraculous recovery because they had never considered such a thing possible. Doctors had always assumed once permanent damage has set in, nothing can be done.

This case and others like it have opened the door for other drugs to be tried in other people in the hope that doctors may manipulate the levels of neurotransmitters involved in the modulation of consciousness and perhaps "awaken" people who have been living as husks of their former selves. Although this line of investigation is still in its infancy, several successful cases of patients treated with certain drugs are now beginning to come to light and new research is under way.

The *60 Minutes* report also reviewed the case of George Melendez, another man who entered a minimally conscious state after nearly drowning, which is another example of anoxic brain injury. Melendez's wife was told by doctors that he would never recover from the brain damage he sustained. She moved him to their house to care for him. Melendez's baseline state was essentially that he would make frequent incomprehensible and loud moaning sounds, which were particularly troubling at night. His persistent moaning would keep her awake frequently, and so one night, sleepless and exhausted, she decided to give George a dose of the sleeping pill Ambien (zolpidem) through his feeding tube, in the hope it would lessen his moaning. As expected, he quieted down. She then entered his room a short while later to check on him, but instead of finding him completely sedated, she found quite the opposite: George was wide awake, alert, and could even talk to her!

She was shocked and couldn't explain what was going on. She was told that his condition would be permanent, but against all the odds, her husband had awakened years after his consciousness had seemed lost forever. Then as quickly as it had come, a few hours later his consciousness disappeared again and Melendez went back to being his old self again—only able to moan and groan but with no other visible signs of being aware or conscious. Unable to explain or understand what was going on, Melendez's wife suspected it must have been related to the sleeping pill. She proceeded to give him another dose of Ambien, and the same thing happened again. For the first time in years, he had become alert and began to answer basic questions again. The Ambien became a daily routine that maintained Melendez's consciousness.

Schiff conducted a brain scan on Melendez to study exactly how the Ambien was affecting his brain. Under the scanner, the front lobe of the brain was yellow without any Ambien in Melendez's system, indicating greatly reduced activity. After Melendez was given the Ambien, the frontal lobe of the brain (front part of the brain) became bright red, indicating that the metabolic activity in his brain had increased by two- to threefold. It was clear that Ambien somehow made Melendez's frontal brain regions, which deal with the modulation of consciousness, come alive (the brain circuit that deals with consciousness includes, among other structures, the frontal sections of the brain). It appears that in Melendez's case the main impairment to being conscious and aware was damage to the frontal regions, which would be transiently corrected with the administration of the sleeping pill; when the effect of the pill wore off, his consciousness would also disappear. Through the use of Ambien, Melendez, like Herbert, progressed from being minimally conscious to being aware again, albeit severely disabled. So in Herbert's case, the anoxic brain injury had caused a deficit in dopamine levels, whereas in Melendez's case, anoxic brain injury had caused a deficit in the functioning of the frontal portion of his brain. Either way, the result had been that each man's consciousness, self, or soul had disappeared until serendipity had made it reappear.

Clinical trials are now under way to study the so-called Ambien awakenings. Researchers attribute Ambien's effect on awakened patients to a phenomenon called paradoxical excitation, the condition that has linked Ambien to sleepwalking, sleep eating, and sleep driving. Though there has been no definitive conclusion about which patients the drugs will help, it is clear that a number of people who were declared to be in a vegetative state (with no chance of recovery and completely absent consciousness) can now be categorized as being in a minimally conscious state or even fully conscious. To ascertain the difference, the patients must be examined repeatedly through the day over a period of time.

"The one liner you get [in medical school] about brain injury was damage done," Schiff said in the *60 Minutes* interview. "We know enough now to know that there are some minimally conscious-state patients where that statement is false."

Approximately two hundred thousand people in the United States are classified to be in a vegetative state, but several researchers now believe as high as 40 percent of those may need to be rediagnosed. Traditionally, a person was declared to be in a permanent vegetative state after three months of brain inactivity for injury resulting from oxygen deprivation and after twelve months for injury caused by trauma. But recent Ambien awakenings and other similar cases are causing doctors and researchers to reclassify many patients in a vegetative state as actually being in a state where evidence of consciousness and awareness can be found even though it had been thought to have been lost forever.

Since the Ambien awakenings research, many other drugs have been found to impact people who seemingly appear to have no consciousness present by altering some specific component of the overall circuitry involved with the modulation of consciousness—to make consciousness "reappear." The difficulty is understanding what specific neurotransmitter or area of the brain circuit needs to be manipulated in each case. This is the most challenging part, since while some people respond to these drugs, others do not because the injury may have affected some other area that is not obvious to researchers.

In fact, these newly discovered cases resemble earlier similar cases also discovered by serendipity. In his landmark book *Awakenings,* Dr. Oliver Sacks recounted the remarkable story of a group of patients who contracted sleeping sickness during an epidemic after World War I. Frozen for decades in trancelike states, these men and women were written off as permanently "gone" due to brain damage. In 1969, Sacks gave them a new drug called L-dopa. The drug had an astonishing and explosive "awakening" effect. It activated their dopamine pathways, and their dormant consciousness was brought out. It all makes sense now, since dopamine is among the many neurotransmitters involved in the modulation of consciousness.

Dr. Adrian Owen, a prominent British neuroscientist, believes that new technology may help doctors properly diagnose more cases. In 2006, in one of the most pivotal and groundbreaking studies, which put to question much of the previously held paradigm regarding minimally conscious and vegetative states, Owen and his colleagues at the University of Cambridge asked people in a persistent vegetative state, as well as normal people in a control group, to imagine they were playing tennis. Owen performed brain scans on the people as they pictured themselves rallying on the Centre Court at Wimbledon. While expecting to see no changes in the brains of people suffering with persistent vegetative states (since they had by definition been considered not to have any consciousness or awareness and had not shown any visible external signs of consciousness in the past), astonishingly the resulting brain images showed that the motor cortex (part of the brain that modulates movement) was activated in almost the same manner in those people who were in a vegetative state as those who had normal functioning brains. Those in vegetative states had completely understood the instructions being given to them by the researchers and could imagine themselves playing tennis just like healthy people. In effect, contrary to everyone's beliefs, their consciousness had always been present, even though appearing to be gone forever. However, it was only through the advent of new research techniques and technologies that doctors had managed to identify subtle evidence that these individuals' consciousness, self, or soul was still there somewhere. More important, consciousness, although seemingly absent, had been present and had never actually been "lost" in the true sense of the meaning—even after extensive and permanent damage to the areas of the brain that modulate it.

Owen, who has since moved his research team to the University of Western Ontario in Canada, conducted another study in 2011 that found brain activity in patients who had been declared in a persistent vegetative state. Using an EEG, Owen and his team measured brain activity in sixteen people in vegetative states and twelve healthy people in response to verbal commands to wiggle their toes or make a fist with their right hands. Remarkably, three people in the vegetative state showed activity in the area of the brain that plans body movement, again confirming that the entity of consciousness, self, or soul was present and had not been lost forever. It was just not visible to the outside world and would never have been discoverable if it hadn't been for the progress in scientific methods and techniques.

How can this be? These were people who had been declared in persistent vegetative states and in whom consciousness was no longer present. The common belief, even among doctors today, is that when someone has a persistent vegetative state due to brain damage such as occurs after anoxic brain injury, there is no consciousness remaining and it will never return. Yet in the case of Melendez and Herbert, the numerous other new case reports of "awakenings" (using various drugs), and Dr. Owen's cases, the self or soul and consciousness had been present even if seemingly lost to the outside world.

These new findings have started to challenge even the most strongly held opinions and beliefs regarding what happens to consciousness in people after suffering the effects of anoxic injury states. Consciousness is not lost, and the brain may be manipulated enough to reveal the hidden consciousness even in people assumed to be in vegetative states and without any consciousness many years after it was thought lost. This suggests that the brain and consciousness are linked, but an even bigger question is, What can we surmise regarding what happens to consciousness, the psyche, or the soul after anoxic brain injury when the heart has stopped? What happens to the human mind and consciousness when we die?*

As WE HAVE SEEN, death is by definition the state that follows after the heart has stopped beating, and there is thus no blood flow or oxygen being delivered to the organs in the body, including the brain. When a person dies, there is no visible sign of life. All bodily functioning, and in particular brain functioning, ceases within seconds of the heart stopping. There is thus no heartbeat, no respirations, and the pupils of the eye become fixed and dilated (due to a lack of blood flow to the brain).

But the big question is, Does consciousness, the self, or the soul become lost, in the true sense of the word, immediately after death? Does it become annihilated forever as an entity at that point?

The answer that is coming out more and more seems to suggest that it doesn't. Consciousness or the soul, while down and thus invisible to the outside world, is not lost forever as an entity—just as we have seen in anoxic brain injury, which is what is happening after death and causes progressive cell damage and death in the brain over time. So while the circuits are clearly down, it is most likely that as we have seen with people who have experienced transient loss of consciousness due to the effects of drugs or any severe medical illness on the brain circuit that modulates consciousness, or even those who have suffered permanent long-term damage to those brain circuits, consciousness is not "lost" in the true sense of the word. In fact, we can study what happens to consciousness in

---

*We can ask these questions because, biologically speaking, anoxic brain injury leading to a stroke is the same as what happens after death. For instance, the main difference between a case such as Herbert's and someone who dies is that Herbert had been saved before his heart had stopped, but the effect of anoxic brain injury on the areas of the brain that modulate consciousness would be similar.

people even after permanent brain damage to the circuits that modulate consciousness (and other parts of the brain) from anoxic brain
injury has taken place, but where the heartbeat has been preserved
(such as Herbert, Melendez, and some of the patients described in
Adrian Owen's study who were suffering with persistent vegetative
state but remained alive).

Clearly from a scientific perspective, we can't answer the ultimate question of whether or not eternal permanency with respect to
consciousness, the self, or the soul continues, not least of all because
today we don't have the means to directly measure and detect consciousness and we thus cannot test such a possibility. What we can
say, however, is that at least for the first few hours after death, which
is the time where we can study things today and also bring a person
back to life, the mind, consciousness, psyche, or soul—whatever
term we wish to use for the "self"—continues to exist. In the cases
where we manage to reverse the process of death and resuscitate the
person "back to life," even many hours after death, the person's consciousness, self, or soul will also come back. So by definition there
has to be some sort of "afterlife," even if only for a few hours after
death. We cannot comment with certainty beyond that point now,
since that is the longest period today in which we can reverse death
and bring someone back to life, but the idea that when we die there
is nothing that remains seems at best premature.

THE ULTIMATE QUESTION AS to how long consciousness (or soul)
continues after death can only be definitively answered when science has discovered a scanning machine (like a brain scanner) that
can detect the entity of human thought and consciousness. With
such a machine it may be possible to continue to track what happens to a person's consciousness (or soul) for longer periods of time
directly after death. While it sounds like science fiction at this point,
and is even incorporated in elaborate stories found in novels such as

Michael Cordy's *The Lucifer Code,* I would not be surprised if scientists do eventually manage to discover a type of scanner that can detect and measure what we call human consciousness. This will help us not only understand what happens when we die but also enable us to better treat people who are living seemingly without their soul or consciousness present, such as we have seen in the cases of people suffering with persistent vegetative states. It will also help us understand once and for all how thoughts, consciousness, psyche, or soul are related to the brain.

In the meantime, when we examine the evidence accumulated so far from resuscitation science through studies of the brain during and after cardiac arrest, objectively speaking we have to at least consider the possibility that the human mind and consciousness could be a separate, undiscovered scientific entity that is not produced by the brain. However, it interacts with the brain and thus continues to exist after biological death has started. Evidence is mounting that those who are brought back from death through resuscitation techniques can tell us what they experienced as well as specific details relating to their own periods of resuscitation—such as conversations or actual events that took place—even though the brain was in a state in which it was unable to function.

This also explains what has long been called a near-death experience, which, as discussed, is better renamed an actual-death experience, at least in the context of people who have suffered a cardiac arrest, since they are not near death but have actually died.

Clearly, understanding the nature of consciousness, the psyche, or the soul would be nothing short of revolutionary for philosophy, science, and medicine—and for humanity as a whole. However, until the direct discovery of the nature of consciousness and its relationship with the brain, one way to go forward would be to study consciousness during the period after the heart has stopped and a person has gone through death, but before they are resuscitated back to life. This would allow us to determine whether objective evidence regarding the continuation of consciousness can be found. This is what ultimately led a group of scientists and researchers to try and undertake such a study, AWARE (AWAreness during REsuscitation), which is aimed at studying the brain and consciousness during cardiac arrest to discover through science more about what really happens when we die.

CHAPTER 10

The AWARE Study

SEPTEMBER 11, 2008, WAS an important day in New York. It was hard to escape the somber memory of the devastating events that had shaken the city and the world seven years earlier. While many commemorations were taking place at Ground Zero, where the lives of all those who had passed away were being celebrated, another important event was also taking place not far from Ground Zero. A symposium entitled "Beyond the Mind-Body Problem: New Paradigms in the Science of Consciousness" was being held at the United Nations Headquarters in New York. The daylong program was sponsored by the NGO Section of the United Nations Department of Economic and Social Affairs (DESA), the Nour Foundation, and the University of Montreal. In many ways, this symposium represented the antithesis of all that had led to the events of 9/11, and it mirrored the mission of the Nour Foundation—to establish a universal platform upon which to draw human beings from all walks of life together in a greater spirit of unity, tolerance, and understanding. The symposium was also an acknowledgment that understanding the complex relationship between mind, brain, and consciousness has widespread ramifications for us all, regardless of our color, creed, or gender. And along those same lines, it made clear that at the heart of the matter, at the core of all human beings and all human actions, lies human thought and consciousness—however ethically or morally developed—and it is this thought that shapes all human ideas and actions, from the destruction witnessed on 9/11 to the solidarity, generosity, and unity that ensued in the following weeks and months.

Among a number of distinguished speakers, I had been invited to speak in front of an international audience, and this opportunity was both a tremendous honor and the culmination of years of work and thought—yet it was also just the beginning. During this symposium, whose main focus was the nature of the "self" and the phenomenon of consciousness, we were also announcing the launch of the AWARE study. The AWARE study—AWAreness during REsuscitation—would be an international collaboration of scientists and physicians joining forces to study the human brain and consciousness during clinical death. It reflected my professional interest and work, which has been driven by two parallel avenues of investigation that are intricately linked. The first is improving the quality of resuscitation in order to save more lives and prevent brain damage. The second is understanding what happens to the human mind and consciousness during death, for cardiac arrest is perhaps the only circumstance in which we can study human consciousness at a time when the brain has naturally shut down and flatlined. This thus enables us to determine its relationship with brain function, while also not losing sight of the fact that when trying to save a life, we are not dealing with a medical process alone but rather a human being. Even if we can't see the human mind, consciousness, or soul, because the neural circuitry that modulates consciousness and awareness is down, a person is in there somewhere, and we must never ignore or forget that.

After successfully completing an eighteen-month pilot phase at selected hospitals in the United Kingdom, which had been established with the help of my colleagues Dr. Peter Fenwick and Ken Spearpoint, we were expanding the study through other medical centers in the United Kingdom, Europe, and the United States.

FOR YEARS, EVIDENCE HAD been building that people who had been resuscitated and brought back from death recalled certain experiences, as well as in many instances specific details of what had happened to them while watching doctors and nurses working on them from above. Could these so-called actual-death and out-of-body experiences be real? Was this why people claimed to be able to see events taking place below while watching from above? Alternatively, perhaps these experiences were simply a reflection of better-quality resuscitation of the brain? Maybe some people were able to recall specific memories because physicians had achieved better blood and oxygen delivery to the brain without realizing it. This was a possibility that had to be considered. Although previous research had indicated that, in general, during cardiac arrest, doctors cannot get enough blood into the patient's brain, there was no way to rule out the possibility that some individuals who had recalled specific experiences had not somehow received better resuscitation. Maybe there were exceptions to the rule. This would be highly significant because it would indicate a potential avenue for further scientific exploration in the quest to identify improved quality brain resuscitation. At that time, no specific real-time brain monitoring system had been identified that could provide physicians with information regarding the quality of oxygen delivery to the brain during CPR, which was vitally important in the quest to reduce brain injury and improve overall survival. The first line of thought was to therefore try and identify a system that could remedy this very important and significant deficiency in our system of CPR care that had been highlighted by survivors with ADEs. Clearly, if we could identify a marker, some sort of gauge and tool that would give doctors real-time second-by-second immediate feedback regarding the quality of their resuscitation efforts and its impact on the brain during CPR, this would potentially help save many more lives and also prevent people from suffering with brain injury. Doctors would be able to potentially quickly recognize circumstances in which oxygen delivery to the brain had been inadequate and then attempt to remedy the situation before it became too late. An additional benefit of such a system would be to provide doctors with the ability to recognize when their resuscitation efforts may be futile and hence when they need to stop CPR, since the call to stop resuscitation in general is somewhat subjective. If after the recognition of poor-quality oxygen delivery to the brain, satisfactory oxygen delivery cannot be established in spite of all efforts, then continuation of CPR would clearly be futile. Thus, a major thrust of the study was focused on identifying a novel mechanism that would enable physicians to evaluate the quality of the resuscitation of the brain during CPR, with the hope that identifying a gauge that reflected quality would ultimately lead to improved resuscitation methods.

Although at first glance, many doctors, including me, would have considered the experiences recalled by people during their period of death to potentially reflect some sort of hallucination, this no longer seemed such a viable explanation. By the time my colleagues and I began the AWARE study, we had come to realize that millions of people all over the world had reported so-called ADEs, including many anecdotal reports of the ability to see and hear accurately during cardiac arrest and resuscitation. This suggested that this was an area we had to investigate further, since ultimately we as physicians are dealing with human beings with real lives, who also have unique mental and cognitive lives, rather than just numbers and statistics. From a scientific perspective, it was becoming clear that it was unlikely the entity we refer to as human consciousness is "lost" in the true sense of the word in the initial phase after death has taken place. Human consciousness could potentially continue to exist in some capacity at least for some time after death, even though it may disappear immediately from sight with respect to the outside world after the heart stops. This is because no (or inadequate) blood flow is getting into the brain during cardiac arrest and the ensuing resuscitation process, which is insufficient to enable brain circuits to function (including the circuits that modulate consciousness and awareness of the outside world). Perhaps these ADE reports were in some way similar to the recently discovered reports of consciousness in people who had been suffering with persistent vegetative states. They too had been deemed not to have any consciousness at all, since the areas that modulate consciousness in their brains had been irreversibly damaged. Yet studies had now demonstrated the entity we call consciousness had not been lost forever in this group of patients in spite of extensive brain damage. It had simply appeared to be "out of sight." Similarly, perhaps people who could also demonstrate conscious awareness even after the process of death had started were the same, particularly since biologically speaking, death shares the same biology as those who have suffered with terrible and progressive irreversible anoxic brain injury.

The evidence so far seems to suggest that the occurrence of consciousness in relation to cardiac arrest is somewhat of a scientific paradox that cannot be easily explained using our current neuroscientific models. This is because consciousness (or the soul) seems to continue to exist and function during cardiac arrest and death, as evidenced by the ability of people to have well-structured thought processes, complete with reasoning and memory recall, when the brain circuits that modulate consciousness are down and consciousness appears lost to the outside world. People can recall specific conversations, details, and events that could only be done with a normally functioning brain. Throughout the years that I have been working in this field, I have come across numerous examples recalled by various physicians who have resuscitated patients back to life.

In September 2012, I was invited to give a lecture regarding the subject of near-death experiences at a conference entitled "Emergency Cardiovascular Care Update," which focused on the topic of cardiac arrest. At the end of my talk, Dr. Tom Aufderheide, a prominent figure in the field of resuscitation science who had been sitting in the audience, volunteered to tell the audience about his own recollection of the first patient he had resuscitated back to life as a new medical intern. He said:

> *I was a brand-new doctor. . . . I had in fact been a doctor for just five days and had never treated a patient with a cardiac arrest. I was told [by my superiors] to go and see a patient who was having a heart attack on the CCU. I walked into the room and introduced myself, and the gentleman introduced himself back. Then at that point his eyes suddenly rolled back in his head, and he fell back into his bed. Being a doctor for just five days, I figured there were probably only two options to account for what had just happened—either he had fainted, or he had suffered a cardiac arrest. I knew it was the latter, as I suddenly saw five nurses run into the room with terrified faces! At that moment my own worst fears had been realized. I was all alone. I had no one to collaborate with, and I had never taken care of a cardiac arrest patient before. A thought directed to my seniors who had sent me to the room alone rushed through my head: "How could you do this to me?"*
>
> *But I got over that really quickly and started CPR. In those days there was no cath lab. There was no therapy for a heart attack. You would just leave the person to finish his heart attack, and if he had a cardiac arrest you would shock him quickly [give an electrical shock using a defibrillator]. Finally after ten minutes of CPR, many more people came into the room, but he just kept on rearresting [having cardiac arrests]. This process went on for quite some time, and the doctors who were in the room had other things to attend to—so what did they do? They left the intern to stand by and deliver the shock treatment when he needed it again. So I remained at this man's bedside*

from 5:00 A.M. to 1:00 P.M. in the afternoon, shocking him repeatedly when he went into ventricular fibrillation. He had a prolonged cardiac arrest. At this point the housekeeping staff came into his room to serve his lunch. I was hungry. So I ate his lunch! I certainly couldn't leave his room, and he wasn't going to eat it!

We finally stabilized him after many hours, and he ended up having a long and complicated hospital course. Then some thirty days later, on his last day before discharge, he said to me, "Can you please shut the door and come and sit down?" I thought that was kind of funny, so I went and shut the door and sat down. He said, "I want to tell you something. I have been meaning to tell someone, and you are really my doctor. You have been here the most, and I felt I can share this with you." He then went on to describe a complete near-death experience. He went down a tunnel. He saw the light. He talked to his dead relatives. He talked to a higher being and was ultimately told he needed to come back. This was a really detailed and prolonged near-death experience, but at the end of it he said, "You know, I thought it was awfully funny . . . here I was dying in front of you, and you were thinking to yourself, 'How could you do this to me?' And then you ate my lunch!"

So that certainly got my attention in the first five days of being a physician! I have been fascinated by the experience ever since, and I often ask my patients about their experiences. It seems to be recalled by about 10 percent of them.

Later on somebody in the audience asked him whether he had vocalized or expressed to the nurses his thoughts, fears, and frustrations about being left alone by his seniors to deal with this very complicated medical emergency. He said, "No, I just thought it to myself and hadn't said a word to anyone else. The thought just glanced past my mind for an instant."

During the question-and-answer session after my lecture, Edward Stapleton, another prominent resuscitation expert who had a background working as an ambulance paramedic, also told everyone in the audience about a person he had resuscitated who had shared his own detailed experience with him afterward. These cases were similar to others recounted to me by other physician colleagues in the United Kingdom, including Dr. Douglas Chamberlain and Dr. Richard Mansfield.* The one common feature among all these accounts was that patients with cardiac arrests had come back and recalled incredibly detailed accounts of conversations and events relating to the period when they were seemingly "dead" to their physicians. Specifically, they all claimed to have been able to see the events relating to their own cardiac arrests while watching from a point above at the ceiling, and their physicians agreed with their accounts, yet found them scientifically inexplicable.

Thus, the key to determining what was really happening would be to have some sort of impartial yet objective test to determine whether or not consciousness could really be present and whether in particular the visual recollections that were being reported were really happening, and if so, when they were happening. Was it during the period of cardiac arrest and resuscitation or some time later when perhaps the brain had started to come back online after people had survived? To objectively test the claims that people had during out-of-body experiences and based on an adaptation of a previous study, we installed images on the top of shelves specially attached to the wall near the ceiling by us. The shelves were the size of a piece of copy paper and were attached approximately six feet, five inches above the ground so that an image that was placed on the surface of the shelf could only be seen by someone looking down from above and not by anyone looking upward from the ground level. The idea was simple: if these recollections and reports of consciousness of being at the ceiling and looking down were cor-

---

*These cases have been described in detail in my book What Happens When We Die? (Hay House, 2008).

rect, then people should also be able to see the images that had been placed near the ceiling. If they were not, then they should not be able to see the images. So if we had one hundred or two hundred people who all claimed to be able to see themselves and doctors and nurses working on them from above and they all saw these pictures, then their experiences would potentially have to be considered real, whereas if none of them saw the images, then they would have to be considered as being less likely to be real—and perhaps were an illusion that had formed after the brain had recovered. Such a study would of course have many obstacles, but at least it was an idea worth pursuing.

In the study, in addition to investigating consciousness and recollections, we sought to examine the processes involved with optimization of brain and organ resuscitation in order to save brains and lives and to study mental and cognitive processes, realizing they are closely linked. It was clear that there is a thinking, conscious person inside the body, and this is what we are trying to save even if the person seems to be absent when we are working actively on them. We would thus study the recollections of those who had survived and would also assess the impact of the quality of resuscitation care, to try to determine if consciousness was simply a reflection of differences in the quality of care with respect to the brain or something more.

To measure the changes in oxygen levels in the brain during cardiac arrest, we obtained a highly sophisticated brain monitoring device, called a cerebral oximeter, that could measure oxygen levels in the brain continuously and record the levels every few seconds during a cardiac arrest and allow us to see if any relationship exists. Cerebral oximetry is a noninvasive system that can be placed on the forehead to give doctors a clear indication of how much oxygen is getting into the brain during and after resuscitation. By measuring the levels of oxygen in the brain during resuscitation of those people who survived and had experiences, we could determine if they were related to the quality of resuscitation as well as identifying a real-time system to inform doctors of the quality of care being delivered to the brain. Although cerebral oximetry had been around since the mid-1990s, it had not been used to track the quality of the delivery of oxygen to the brain during CPR. There had been a few occasional instances in which it had been used during CPR, particularly in the case of certain people who had suffered an unexpected cardiac arrest while being monitored using this technology during surgery for another purpose. It had shown promise but had not been systematically investigated during cardiac arrest resuscitation. Perhaps the people who had experiences simply had higher levels of blood flowing into the brain and hence oxygen delivery. Either way, by studying this technology in a novel manner, we would potentially be able to change the way that doctors manage cardiac arrest by developing a new application for this already existent technology during cardiac arrest and hence ultimately reduce brain injury.

Such an endeavor aimed at studying people's experiences from a period of cardiac arrest would of course have huge challenges. The biggest challenge would be that, unlike other research projects where researchers could enroll patients from clinics or elsewhere, we were dealing with cardiac arrest, an event that can happen randomly at any time of the day and anywhere in a given hospital. The second major challenge (as I have explained) was that most patients who suffer a cardiac arrest by virtue of having died once are unlikely to survive long enough to be interviewed so that we can determine if they had any experiences or not. So we would be dealing with an event that could happen anytime, anywhere and would give us a roughly 15 to 18 percent chance of survival to a point where the person could even be interviewed. Of course, many of the 15 to 18 percent of people who survive will end up with permanent neurological and cognitive damage and so would therefore not be in a physical state to be interviewed. On top of all these challenges, we would have to account for the fact that most of the people who do survive and are in a state to speak will of course have memory loss resulting from the cardiac arrest itself and the impact of all the changes in the brain that take place both during cardiac arrest and in the postresuscitation period. Finally, we would have to address the question of why people having out-of-body experiences (assuming they are even real) would even look at the images we put up (rather than focusing on the doctors working on them).

So what type of images should we choose, and where should they be placed? The ideal position would be on the patient's bed, but of course, that would mean that everyone would see them; we couldn't then be sure that, if a patient recalled the images, it hadn't been because somebody else who had been part of the resuscitation had told the patient about the images after the event. We decided to place the images directly above the head of the bed, just below the ceiling so that they would be in the line of vision of patients looking at themselves. We would also use images that would be potentially interesting and eye-catching to people. For example, if we felt the people were very nationalistic, we could use symbols that reflected their country, whereas if they were religious, we could use imagery that reflected those beliefs, and so on.

Our initial goal was to recruit twenty-five hospitals with the objective of recruiting at least fifteen hundred cardiac arrest survivors. By 2008, we had reached agreement with investigators from twenty-five different hospitals to take part. Large numbers of hospitals were needed because the low survival rates meant that we would need ten thousand actual cardiac arrest events to have fifteen hundred survivors. But then again, at best 10 percent would be expected to have any memories (ADEs), and only 2 percent would be expected to have an out-of-body experience that would enable us to determine if what they claimed to see was correct. Although many people have an ineffable experience that includes feeling peaceful, seeing a luminous being that guides them, or entering a beautiful place, these experiences were very subjective and could not be independently tested. The only component that would be amenable to testing would be an out-of-body experience, because it was only during this time that people recalled specific potentially verifiable events that had taken place relating to their own cardiac arrests. Based on these numbers, if we followed ten thousand cardiac arrest events, we would perhaps expect to end up with about 150 people with mental and cognitive experiences and perhaps only thirty people with an out-of-body experience.

Reflecting on that day in 2008, it was clear that many professional factors had led to the announcement of the AWARE study. On a personal level, my speech at the United Nations returned me to events during my own childhood and early adult life where I experienced firsthand what it means to live with someone with severe brain injury. This early life experience had brought home and made very clear to me the importance of trying to save "brains" while studying the nature of human consciousness. When I was nine years old, my father, then thirty-seven, was diagnosed with a progressive and devastating neurological disorder. In eighteen short months, he was in a wheelchair, and the crippling effects of this neurological disorder continued relentlessly until he became figuratively like a "vegetable" within just a few years. He endured this condition for seventeen years, and for the majority of that time, he was just barely alive and unable to communicate at all. As a teenager I would spend time with him, not understanding what had happened to my father, the man who had been so strong and powerful in my eyes when I was a small boy, but now had become just a shell, a husk of his former self. We couldn't communicate in any way other than being in the same room together, and he couldn't reminisce about the things we had done together or even enjoy life's simplest pleasures. Later, when I graduated from medical school, I couldn't even tell if he was aware that I had become a doctor—something he had always wanted me to aspire to when I was a young boy. He was trapped in a body that could no longer convey his wishes, his feelings, or his thoughts to the outside world. We all felt he was there—but often wondered where and what was really left of his self. As he lay propped up by cushions in his bed day after day, I often wondered what was happening to this man's consciousness, to his sense of self—that entity that made up his unique personality. As sad as death is, it was a great relief for all of us when he finally passed away at the age of fifty-four, because he had endured such suffering.

During those seventeen years, I witnessed up close and personal the devastating effects of a destructive brain disease. The end result of my father's neurological condition and his state of being was largely the same as what happens to people who suffer catastrophic brain damage after anoxic brain injury from cardiac arrest. In truth, part of what motivates my work and drives me is the realization that through the establishment of higher standards of care, we can all work together in unity to ensure that many of the people who have suffered cardiac arrests and have been resuscitated not only come back to life again but also do not end up neurologically in a state similar to Don Herbert, George Melendez, or even my own father. Professionally, I've resuscitated hundreds of people and taken care of hundreds more who could not be revived. I also unfortunately have seen people on the other end—those who have been resuscitated and ended up with the devastating consequences of brain injuries after cardiac arrests. It saddens me to feel that people may be dying or getting brain damaged unnecessarily, because with better systems many cases could potentially be avoided. This, of course (as I have tried to highlight in this book), is a complex problem that requires changing old perceptions while introducing a systemwide change regarding how medical treatment in this area is administered both in the community and in hospitals, much like what has been done in the aviation industry.

At the time of launching AWARE, my colleagues and I had spent nearly ten years searching for the funding required to enable us to perform this study on a large scale. It was clear that this work was needed now, if only to enable us to design better studies. If we didn't take the first step, then we wouldn't be able to take the second and the third. The study received funding from the Resuscitation Council in the United Kingdom and later additional support from the Nour Foundation to develop the study in the United States. For the European arm of the study, we also received funding from the Bial Foundation, a Portuguese nonprofit organization. Although ideally we wanted to have fully funded positions of independent staff in each hospital doing the work, that would prove far too costly when considered in the context of twenty-five hospitals, and it would be too complicated since no one could predict when or where a cardiac arrest would take place. Most cardiac arrests actually happen at nights and on weekends, so even funding the position would not guarantee having someone there when we needed that person the most. We knew we needed to start somewhere. So we agreed that staff at the twenty-five hospitals working with us would use their own time to identify cardiac arrest patients, interview them, and study their charts for various markers that would show the quality of care they received; staff would then forward the data to us. We realized we would probably capture fewer people that way, but at least it was a practical way to go forward. Because of the enormity of the project at hand and the number of beds at each hospital (average 500 beds, multiplied by 25 would total roughly 12,500), combined with funding constraints, we also couldn't install a shelf in every room in every hospital, so we worked with our investigators to determine historically where patients were more likely to have cardiac arrests, such as in the emergency room or the coronary care unit. We knew there would be cases where someone would have a cardiac arrest in a room without a shelf, but again for the sake of being practical and ensuring the study would be manageable, we had to compromise. Clearly, although the study would be more complete with more resources, we needed to take this first step. In all, we installed about one thousand shelves to begin with, which was itself quite a lot, but still covered only 10 percent of the total hospital beds.

Obviously, we didn't know what to expect at the time, but nevertheless, we proceeded with the study.

NOW IN SEPTEMBER 2012, it has been four years since we started our research collaboration. This period has enabled us to test many of our systems with a view to building on our findings.

The results so far have been very promising and have demonstrated that almost all the patients in our study who have suffered with cardiac arrests start out with very low brain oxygen levels. Normally brain oxygen levels should be between 60 percent and 80 percent; however, in our experience during cardiac arrest, the levels are often well below 20 percent and on many occasions below 10 percent and even sometimes zero! Importantly, we have found that if quality resuscitation efforts are instigated during this time, the level of oxygen to the brain can gradually be increased with sustained effort. Nevertheless, it has become very clear that if the levels are less than 30 percent, the heart will almost never restart and the person cannot be revived. On the other hand, when doctors have successfully elevated the oxygen level through quality resuscitation efforts in the brain to approximately 45 to 50 percent and maintained these levels for approximately five minutes, then in almost all instances the heart will restart. The key is thus to elevate the oxygen levels in the brain to at least this important threshold level.

This was a significant yet unexpected finding, and it led to a very interesting question. What does brain oxygen level have to do with the heart? In other words, if this machine is measuring oxygen delivery to the brain, how is it that it seems to clearly predict whether the heart will restart and the person will be saved, at least in the immediate sense? The answer seems to be that if we can clearly improve oxygen delivery to the brain, then it means that we have also been able to improve oxygen delivery all over the body, including the heart. Oxygen levels as measured by cerebral oximetry thus not only provide data regarding the state of brain resuscitation but also are a surrogate marker of the level of oxygen that is being delivered to the entire body, including the heart. If the levels that are being delivered to the heart are below a certain threshold (30 percent), then the heart will simply not have sufficient oxygen to be able to allow the heart muscles to pump again, and the person remains dead and cannot be revived.

Although this early data is extremely important because it helps us predict when immediate survival can be achieved, it doesn't tell us the brain oxygen levels that may predict longer-term survival and also reduced brain injury. At this point we are thus concentrating our efforts on trying to identify the optimal oxygen level in the brain that would potentially lead to better longer-term survival and also reduced long-term brain injury. It is clear that while a level of about 45 to 50 percent may perhaps be sufficient to restart the heart in the immediate phase, it may not be sufficient to reduce the post–cardiac-arrest tsunami effect that inflames the brain and other organs with a toxic fury leading to brain swelling and generalized organ injury after the heart has been restarted and the patient has been revived initially. This is the main reason people die in the subsequent hours to days. Our preliminary data suggest that we may have to strive to achieve a higher level of brain oxygen level to not only start the heart but also, and more important, preserve the brain and minimize the toxic tsunami effect that ensues when there has been a prolonged period of lack of oxygen to the brain during cardiac arrest.

Interestingly, roughly at the same time that we started doing this work, another group (Dr. Noritashi Ito and his colleagues at Osaka Saiseikai Senri Hospital) in Japan had also started to study the role of cerebral oximetry during cardiac arrest. They specifically looked at this technology in patients who had been taken to their hospitals by ambulance after suffering with cardiac arrests. Their data indicated that if a patient arrives at the hospital with a particularly low level of brain oxygen (less than 25 percent), then the person would invariably have a poor outcome. The results of our work, however, complement the work carried out by the Japanese group. We have demonstrated that after measuring a low brain oxygen saturation level, specific steps may be taken to actually improve this and hence lead to the restoration of the heartbeat and potentially improve longer-term brain and survival outcomes. Thus, measuring a brain oxygen level that is very low while using a cerebral oximeter as a form of "spot check" does not necessarily lead to poor outcomes, since the oxygen level can be improved with dedicated effort. At this time, we have integrated the routine use of cerebral oximetry into all our cardiac arrests at my own particular medical establishment. As far as we are aware, we are one of the first medical centers (if not the first) to do so. We have also published the results of our studies in reputable scientific journals and have presented our findings at national and international scientific meetings, including the annual American Heart Association meeting. Interestingly, however, none of the patients in whom cerebral oximetry was used reported any mental and cognitive recollections from their periods of cardiac arrest. Nevertheless, the component of our work that focuses on cognitive and mental experiences and the ADE has also been ongoing and has provided us with a lot of interesting information that we are currently analyzing.

DURING THE FOUR YEARS since the study started, our collaborators in the different medical centers who have been participating in the AWARE study have reported a total of more than four thousand cardiac arrest events. Of these, it was found that on average in approximately 32 percent of cases the heart had been restarted by doctors and nurses resuscitating the patient but that only about 50 percent of these initial survivors had remained alive to a point where they could be interviewed (thus about 16 percent of the total cardiac arrest events), with many unfortunately dying in the subsequent hours to days after their hearts had initially been restarted. As mentioned, during this period we installed in the participating hospitals approximately one thousand shelves high up with images that were only visible from the ceiling. We did not have the means to be able to put a shelf above every single bed in each hospital because of the vast numbers of shelves that would have been required.

During our initial assessment we had estimated that if we covered the most critical areas of the hospitals, we would hope to be able to capture at least 80 percent of all cardiac arrest events, since we know that most cardiac arrest events happen in the same locations in hospitals. These are usually the coronary care units, the emergency wards, as well as specific wards where more critically ill patients are kept, such as the intensive care units.

The evaluation of these initial approximately four thousand cardiac arrest events actually indicated to us that despite all our best efforts, placing the one thousand shelves had enabled us to capture no more than 50 percent of all cardiac arrests in the hospitals. This meant that in at least half of the cases where a cardiac arrest had occurred, it was possible a person who had recalled a specific ADE, and had provided us with recollections of what had actually happened while seemingly watching from the ceiling, was not able to have his or her case tested and examined in a more objective fashion with the use of these image boards.

In this initial phase, a total of approximately a hundred interviews were carried out with cardiac arrest survivors, and we found that ADEs seem to have occurred in only approximately 5 percent of our study group. It was therefore possible that we would have had more patients with specific recollections but they had been lost to follow-up.* Those whose experiences were documented seemed to

---

*The vast majority of cardiac arrest events had been occurring during evenings and weekends at the participating hospitals, and unfortunately the staff who had time to follow the patients were unable to interview all those who had survived prior to being discharged from the hospital. This was mainly because the randomness of the cardiac arrest events as well as their timing, which was usually not during normal working hours, combined with the pressures of the staff's daytime job requirements, meant that a large number of patients had been discharged home before they could be interviewed regarding their recollections.

follow what we had already come to understand of what happens during a cardiac arrest. Although few and far between, the experiences were nevertheless interesting. My colleague Ken Spearpoint described one patient's experience. He wrote:

> *His journey commenced by travelling through a tunnel towards a very strong light, which didn't dazzle him or hurt his eyes. Interestingly, he said that there were other people in the tunnel, whom he did not recognize. When he emerged he described a very beautiful crystal city and I quote "I have seen nothing more beautiful." He said there was a river that ran through. There were many people, without faces, who were washing in the waters. He said that when the people were washing it made their clothes very bright and shiny. He said the people were very beautiful and I asked him if he recalled hearing anything—he said that there was the most beautiful singing, which he described as a choral—as he described this he was very powerfully moved to tears. His next recollection was looking up at a doctor doing chest compressions!*
>
> *For the patient this was a profound spiritual experience, and certainly powerful for me too . . . unfortunately the event was not in a research area [an area with a board].*

For a long time, nobody in the first phase of the study reported the ability to "see" themselves from the ceiling and watch doctors and nurses working on them. We had one "near miss" in 2009 when everyone thought they had had a "hit," which led to a frenzy of excitement at the participating hospitals in the United Kingdom. A patient who had suffered with a cardiac arrest at St. Peter's Hospital located just outside London had been sent to another participating hospital in London (St. George's) for cardiac testing. The team at St. Peter's had asked the team at St. George's to conduct the interview on their behalf since the patient had left their hospital. During the interview it transpired that the patient had recalled an ADE, and as part of this he had seen a vision of a green light. His experience was not very detailed; however, the staff at St. Peter's were curious. They went back to the ward where he had suffered a cardiac arrest and removed the board that contained the image that had been above his bed. The image on the board was essentially the front cover of a UK newspaper showing smoke coming out of a New York apartment block after a small airplane flown by a famous baseball player had lost control and crashed into the building. At first, it seemed there was nothing to correspond with the patient's recollection, but when they looked more carefully at the bottom of the newspaper cutting, they spotted an image that caught their attention. This was a picture of a group of doctors standing in a circle round an operating room light wearing green scrub caps on their heads. The team thought this corresponded with the green light the patient had recalled. I received a very excited call, but on closer questioning the patient had not recalled seeing anything from above at all.

Then finally in 2011 we had our first real out-of-body experience claim. A fifty-seven-year-old man had suffered a cardiac arrest in the cardiac catheterization laboratory in Southampton General Hospital in the United Kingdom. The patient specifically recalled feeling that he had been above his own body and had been looking down. He said he had seen people in the room around him and that they had given his heart electrical shock treatment (defibrillation) twice. He said he had a bird's-eye view, while looking from above himself, of all that was happening to him below. However, the cardiac arrest did not take place in an area of the hospital where there had been a board present, so the staff were unable to confirm the details of what the patient had recalled; specifically, they couldn't ask him whether he had "seen" an independent objective image. I did, however, interview him later and found his experience to be remarkable. I have transcribed our conversation here (with my questions and comments in italics):

Before I tell you about my experience, I want to tell you about a couple of things that will put it into perspective. In terms of me, I come from a really, very, very large family. I know that directly after I told them about my experience they said, "It was the drugs," and things like that.

After the experience?

Yes, they all said it wasn't real and things like that. But there are a couple of things about it that stuck with me and made me want to repeat [my story], directly to people who would listen to me.

So you felt you wanted to tell people.

Yes, because it was totally alien to whatever I had had before in my life, and I felt it was important and basically it went like this:

Apparently I didn't know I was having a heart attack, doctor. I was at work, I went to work normally, I didn't feel 100 percent, and that is not unusual, because one of the other things I have is diabetes and I have had that for over thirty years now, but I felt a little bit different. Because I felt different, I immediately tested my blood, because when you have had diabetes for a time you get used to things all the time and one of the first things I do is test my blood just to make sure that I haven't eaten something or taken something that I shouldn't have, and my blood was fine—no problem whatsoever. The day went on, I had a coffee and had a meeting, and my boss said, "Let's go down to the pub and we will have a sandwich and a coffee."

What do you do, sir?

I am a social worker.

So do you work for the hospital?

No, I don't. I work for a charity. So that's what I did. We went for lunch. Came back from lunch. We were only gone half an hour and I didn't feel well again. I work in an open office and it's really difficult to . . .

You are visible?

Yes, very visible. You can't talk on the phone privately or anything like that. So I went to the loo, and I felt really unwell. I came back into the office and said to one of the people there, "Can you go and get me a chair, just so I can sit." It felt like I wanted to get some air and I couldn't. So they went and got me a chair and—anyway, cut a long story short—they phoned an ambulance.

What time was this, roughly?

This was about 2:30 in the afternoon. I felt well again, the ambulance eventually arrived, and two paramedics came up. I said, "Look, I feel well now," and they said, "Would you please come down to the ambulance." Halfway down I felt unwell again. Anyway, I got into the ambulance and they put something on me, a wire or something, and then the whole mood changed.

When they got your ECG?

It sort of all changed then, and they wanted to whisk me off and not talk to me and just do it. Do you know what I mean, doctor? That unnerved me a little bit because I am not used to anything like that, so I said, "Hang on, what are you doing?" They said, "We need to get you to hospital." Anyway, they did.

So based on the ECG, they said, "We have to get you into hospital," right? You had changes that said you had had a heart attack. But you didn't have any symptoms.

No pain, no nothing, no pain whatsoever. I can remember coming into the [hospital bay] . . . and a nurse came on board. [The paramedics] had told me a nurse called Sarah would come to meet me when I arrived.

Getting on board the ambulance?

She came on board the ambulance like they said she would and then she said, "Mr. A, I am the most important person in your life at the moment. I am going to ask you some questions and I want you to answer every one of them." I said yes. I can remember that I wanted to sleep all the time at that stage and all she kept trying to do, it felt like, was to keep me awake and talk with her. Do you understand what I mean? And that's how it was with her.

But you were sleepy?

Yes, I felt sleepy then, very much so. Anyway, she was asking me all these questions, have I got pain, have I got this, and I can remember 100 percent. I can remember her asking me one question and because I didn't answer it quick enough or was still tired, she asked me again and I bit back at her, as if to say, don't keep onto me. I can remember doing it. Anyway, they got me into hospital. I can remember lying on a trolley or a bed but I couldn't see any further. I couldn't see what they were doing down below me.

At this point he described that while he was in the catheterization laboratory, the team had placed a sterile drape over him and he was lying flat. The nurses had placed the drape as a form of partition approximately at the level of his upper body so that the doctors and nurses could work around his groin without him being able to see what was happening. He hadn't seen the doctor come into the room, and he was just lying there. Typically the team then gives injections in the groin area to numb the patient and then push a wire firmly into the main blood vessel in the groin and feed it all the way up to the heart.

So you were awake still?

Yes.

You were lying there?

100 percent.

They were getting you ready to do the procedure, the cardiac catheterization?

Yes, to put a stent in.

That's right.

Yes, and she was still talking to me.

Sarah was?

Yes, and I was answering her, but I could also feel a real hard pressure on my groin. I could feel the pressure, couldn't feel the pain or anything like that, just real hard pressure, like someone was really pushing down on me. And I was still talking to her and then all of a sudden, I wasn't.

The medical records from the hospital indicate that at this time Mr. A had gone into cardiac arrest with his heart showing a specific electrical abnormality called ventricular fibrillation (VF). This is a fatal condition in which instead of beating, the heart fibrillates but cannot create a contraction, and hence, there is no heartbeat. The heart immediately stops, and the person receives no blood to the brain and dies immediately. It is a well-known complication of a heart attack and is usually what kills people instantly after the attack. VF is universally incompatible with a beating heart and hence life. There have been many scientific studies that have examined what happens to the brain immediately after the heart goes into VF and stops. These studies have all demonstrated that brain electrical activity stops and the brain itself flatlines.

In the cardiac catheterization laboratory of Southampton General Hospital, the medical and nursing staff had attached a specific type of device that can shock the heart, called an automated external defibrillator (AED), to the patient's chest. This is the same type of defibrillator that is installed in many airports, train stations, and other public places. The difference between this defibrillator and others routinely used in many hospital wards or emergency areas is that the AED is designed to be used by laypeople (but can also be used by health-care professionals), whereas other defibrillators are designed only for professionals. The AED assumes the person using it has no medical knowledge and thus would not be able to recognize VF and know when to deliver the necessary shock treatment. The AED is designed to be very simple to use and can detect VF itself. Once VF is detected, the AED provides verbal feedback to the users. The device would say something like "shock advised" and then enable the delivery of the shock treatment. Standard hospital defibrillators leave the recognition of VF to trained nursing and medical staff and therefore do not have a verbal output. I went on to ask Mr. A what happened.

You blanked out?

I must have done. I didn't know at the time, but then I can remember vividly an automated voice saying, "Shock the patient, shock the patient," and with that, up in that corner of the room [*he pointed to the far corner of the room*], there was a person beckoning me. I can see her now, and I can remember thinking (but not saying) to myself, "I can't get up there." The next second I was up there and I was looking down at me, the nurse Sarah, and another man who had a bald head. I can remember doing that. I can remember seeing them whilst I was up there watching them do that.

What did Sarah look like?

Sarah who?

Sarah the nurse.

She had blond hair. She was quite tall.

And you could see her from . . . ?

Up in the corner, right up in the corner I was, with this other person beside me.

So let's say this is the room. Where were you lying before this happened? Where was your head?

On the bed here, and that's the corner. [*He again pointed to the far corner where his legs would have been pointing had he been lying down. There was a bed in the hospital interview room.*]

Where was Sarah standing relative to the corner?

[Before this happened] I couldn't see down here. [*He showed the end of the bed where his legs and groin area would have been.*] I couldn't see her and I didn't even know there was another man standing there. I hadn't seen him. Not until I went up in that corner—then I saw them. You understand what I am saying?

Before this happened, before you lost consciousness, you didn't see Sarah physically at that point, and you didn't see this man with the bald head?

No.

When you were looking there, were you seeing her front, her face, her back, or what were you seeing?

[*He pointed to the far corner of the room.*] This was me, and she was here. [*He pointed to the end of the bed.*] And the other man was on the other side of me there. [*He pointed to a position next to the nurse where his legs would have been.*]

Which way were they facing?

Towards my head. Away from me while I was looking down from the ceiling.

You could see their backs?

I could see all this side of them. [*He pointed to the back.*] As clear as the day I could see that. [*He pointed to an object.*] The next thing I remember is waking up on that bed. And these are the words that Sarah said to me: "Oh you nodded off then, Mr. A. You are back with us now." Whether she said those words, whether that automated voice really happened, I don't know—only you would know those things. I don't know how to be able to confirm that those things did happen. I am only telling you what happened with me and what I experienced.

In the medical records, the events had been independently documented. It read:

In cath lab recovery area . . . IV access obtained by cardiology registrar. [This is the senior cardiology trainee physician getting access to the blood vessel.]

    Whilst repeat ECG taken—VF arrest. Shock x2 AED 150 joules. ["150" refers to the strength of electricity the AED device had given.]

    Return of circulation 15:07 hours.

May I ask you something else? What else did you see in the room? What I mean is, did you pay attention to anything else or were you fixed on this? What were your feelings? If you remember any? What else if anything were you looking at? And what happened to the lady that you saw?

I don't know what happened to that lady. I can still see her now if I want to. I want to say she was an angel.

Did you recognize her?

No.

What were the distinguishing features about her?

She had lovely curly hair. It wasn't blond but it wasn't dark, if you know what I mean. She just had lovely features about her. I would say she was an angel, but from what I perceive as an angel. I can remember that she beckoned me (I can remember thinking, and I know I didn't say it, "But I can't get up there"), and the next second I was up there.

You thought it?

Yes, 100 percent I thought it.

You thought, "I can't get up there"?

Yes.

So, can I take you back to that thought—when you had that thought, where were you?

In my bed.

Right, but were you physically in the bed or just above yourself? How was it that you saw her? Were your eyes open? Do you know what I mean? How were you summoned?

Whilst I am lying down, when I have looked up there, she beckoned me like this. [*He mimicked being beckoned.*] I can remember thinking, "I can't get up there." You know, whether I said that with my face I don't know, but that is how I felt. I can remember feeling it and the next second, I was up there, looking down on me.

In that split second, before you found yourself up there, did you see anything else around you in the room?

I can't remember.

Okay, it was such a split second that you basically just shot up and you were there. And then when you went there, you didn't notice her anymore—is that correct?

It felt like she was with me.

It felt like there was a presence there.

I didn't say I couldn't see her anymore, because I wasn't looking to my left or right while in the corner. I was looking down at my body lying below.

Did she have any kind of presence? I mean, how would you describe her? Did she have any qualities, any personality?

I felt that she knew me, I felt that I could trust her, and I felt she was there for a reason and I didn't know what that was.

So then, you were basically fixated on what was going on down below—it grabbed your attention?

I was up there looking down at me lying on the bed, and I couldn't see my face because there was like a curtain here, and I didn't know [before this happened] that there was a man on the other side of me [*he pointed to where the drape had been placed that had prevented his seeing what the doctors and nurses were doing to him before his experience*], and I could see Sarah on that side, and that's all I can tell you.

What did the man look like? I mean, what did you see of him? Can you still picture him?

I couldn't see his face but I could see the back of his body. He was quite a chunky fella, he was. He had blue scrubs on, and he had a blue hat, but I could tell he didn't have any hair, because of where the hat was.

What did you see of his head that made you think about it?

Just very little. It looked like he was bald and he just had a hat on. I wear a hat because I like a hat. I am not bald.
    I know who he is.

Who is he?

I don't know his full name, but he is a professor now, and he was the man that I saw later, because the next day, when I was lying in bed on the ward, I saw this man [come to visit me] and I knew who I had seen the day before. I don't know his name, Professor something. He is now a professor—he wasn't at the time, but he is now. [*The hospital staff correctly revealed his name and confirmed his promotion.*]

Because this was 2011, wasn't it?

Yes.

What was Sarah doing?

I don't know what she was doing. All I know is that she was this side of me. [*He pointed to where his legs would have been.*]

What was she wearing?

Blue, well, like that color blue. It was a different blue to what he was wearing. Almost sure of that.

Slightly different blue?

Yes.

What was she doing?

I don't know, something to me, but I don't know exactly what they were doing.

Was she moving her arms?

She was doing some things and she was trying to . . . at one stage, I can remember, her face looking [toward the male doctor], almost in anticipation of his movements, checking to see whether he was going to do something or not. Do you know what I mean? That's how it felt. Then within the next few seconds, I was back in my bed, and she was saying, "You nodded off for a little bit."

What were your feelings like up there? Did you have any sensations, feelings, sentiments, anything?

The only time I realized that my heart had stopped was about twenty minutes later.

No, sorry, I meant when you were up at the ceiling, did you have any feelings then?

I can remember feeling quite euphoric in terms of, I am actually up here, I can see all of this. Do you understand what I mean?

In that split second, and you may have not had time to see or think of anything else, I am just curious, did you notice anything? Because obviously just before that you had this drape or curtain in front of you so you couldn't see around, but then suddenly you were at the corner and you were able to see things. What did you see?

All I could see was a space in front of me and them working on me.

Anything else?

I just focused on myself. But I have thought about it and thought about it for a long time. I just wish that there was a way that I could prove to you what happened, and there isn't a way but I think that I have thought of a couple of things that I think may help.

What would help?

If the doctor, or whoever is doing anything at that stage, picks up an envelope and in an envelope is a word, it could be any word, that is repeated three times between the two people or three people dealing with that procedure—whether I would have picked up on that one word or not, I don't know.

The interview ended soon after, and he gave me permission to reveal his experience and look at his medical chart.

What is most remarkable about this experience is that while this man did not suffer with a prolonged cardiac arrest, the duration of his cardiac arrest would have nevertheless been at least three to five minutes. This is because after recognizing ventricular fibrillation and delivering the first shock treatment to a patient, the AED device will automatically recommend a two-minute period of chest compression before analyzing the heart rhythm again. If the patient is still in VF, the AED will recommend another electrical shock. Therefore the whole process of delivering two shock treatments and the subsequent analysis would take about three to five minutes, which is well beyond the time that it takes for the brain to stop functioning after the heart stops. As described earlier, the brain stops functioning immediately after the heart stops, which is why people lose consciousness instantly (and usually within ten seconds all electrical activity stops and the brain completely flat-lines). This patient, however, maintained conscious awareness of his surroundings and was able to accurately describe events that he had not been aware of before his experience, such as seeing the balding male cardiologist in blue scrubs standing at the foot of the bed. He also correctly described "hearing" the AED give the two separate commands to deliver the shock treatment (which would have been at least two to three minutes apart). This provides some perspective in terms of how long after his cardiac arrest he was able to maintain conscious awareness while perceiving that his real self was at the ceiling. Although there had been no images placed on a shelf in his room, judging by the fact that this man was able to maintain consciousness at a time when his brain could not have been in a functioning state, this case supports the emerging scientific possibility that after death has begun, a person's consciousness, psyche, or soul does not appear to be annihilated. It also supports the position taken by some scientists and physicians (such as professor's Sir John Eccles and Bahram Elahi) that human consciousness (or the soul) could be a separate and independent entity to the brain and may continue to exist after death.

The final practical take-home message from this case is that placing an image above the head of the bed may not have been very useful in this case since he perceived his consciousness as looking from the other corner of the room. This is an important point in guiding any future research that may seek to use preinstalled images to independently test the claim that people are able to "see" from above.

Since that time we have amended the study and sought regulatory approval to contact and interview the patients who had been discharged home from the hospital before they could be interviewed. This aspect of our work is still ongoing, but we hope in the near future to be able to reach out to all the patients from the participating hospitals who are still alive, using a mailed questionnaire followed by a telephone call. So far, from these ongoing interviews we have had one more person claim to have had an out-of-body experience. This occurred in a fifty-one-year-old woman who suffered a brief cardiac arrest in one of the hospital wards. During her interview she had stated:

> *I felt scared. I was on the ceiling looking down. I saw a nurse that I did not know beforehand who I saw after the event. I could see my body and saw everything at once. I saw my blood pressure being taken whilst the doctor was putting something down my throat. I saw a nurse pumping on my chest, saying, "Come on Vanessa, come on Vanessa." I saw blood gases and blood sugar levels being taken, but it did not hurt (it usually does when I am awake). At the beginning, I think, I heard the nurse say, "Dial 444 cardiac arrest."*
>
> *I became depressed after the event and later was frightened to get off the bed. I did not have a chance to talk to the nurse [mentioned above] but I did see her on the ward.*

However, in this case, the cardiac arrest also had taken place in an area of the hospital where there had not been a board present. Not surprisingly, when the interviewer asked if the woman had seen anything unusual such as signs above the bed while she was looking down, she had said no. A board was not present in this particular location because it was one of the wards with a historically very low incidence of cardiac arrest. Therefore, at the beginning of the study it had not been identified as a potential hot spot, and thus no images were ever placed there.

Interestingly and perhaps disappointedly, both out-of-body experiences reported to us so far have occurred in areas without images placed on shelves attached to the walls. Therefore, we have not yet been able to objectively ascertain the accuracy of people's claims to "see" events from above. However, we are still actively conducting interviews, and it is possible there may be more out-of-body experiences reported. Closer examination of the two cases has also demonstrated another very interesting observation. Both cardiac arrest cases had been relatively short in duration (less than ten minutes), which I believe may turn out to be very significant.* As explained previously, if a cardiac arrest event is relatively short, then the postresuscitation inflammation and disease that normally engulf the brain and cause widespread damage (including damage to the memory circuits) are also relatively mild by comparison to someone with a prolonged cardiac arrest. This suggests that people who report profound ADEs, including out-of-body experiences, may perhaps be able to better recall their experiences simply because they had suffered less damage to their brains and specifically the memory circuits in the days and weeks after the cardiac arrest. Maybe many others also experienced an ADE but couldn't recall the details of the experience due to their memories being wiped away by the extensive postresuscitation inflammation and damage that occur in the brain after a cardiac arrest. This may explain why 90 percent of people who survive a cardiac arrest usually say they have no memory of their cardiac arrests, and the remaining 10 percent usually have very sketchy memories with only a very small proportion having detailed recollections. The main finding, though, has been that the out-of-body experience is even rarer than we thought and seems to occur in less than 1 percent of survivors (0.1 percent of total cardiac arrests). This suggests that our original calculations will need to be revised and the study expanded, since after four thousand cardiac arrests, we had only two out-of-body experiences.

Incidentally, this was also the case with the person who had a very profound experience described above.

FOR NOW WE ARE continuing with the data collection process, but once we manage to collect all the initial-phase data for the study, we plan to publish our results in a reputable medical journal and amend and alter the study based on the lessons we have learned. It is clear now that installing images on shelves above beds may not be sufficient, because it takes an enormous effort and in many cases cardiac arrests happen in areas without images. For the moment, the recollection of ADEs is quite rare, particularly when we look at the occurrence of an out-of-body experience. We would therefore need to expand our network of hospitals and, more important, provide funding for each center to dedicate a member of staff who can attend every single cardiac arrest and ensure that all patients at the staff member's site are interviewed within a few days of their initial survival. We will therefore be looking at alternative means, such as the potential use of a tablet computer that may have a built-in image generator and timer device, which a dedicated member of staff at each hospital can take to the scene of a cardiac arrest; the staff person can then place the image generator at a point high above the patient's head during cardiac arrest. The images would rotate sequentially and the data could hopefully be captured in one go and then be downloaded and analyzed by an independent member of staff.

Today, we are also continuing our efforts with respect to the use of cerebral oximetry, both in terms of the relationship that it may have with patients' experiences and, more important, as a unique way to determine the quality of resuscitation; we hope to identify novel ways to improve oxygen delivery to the brain with a view to ensuring more people survive and have less brain injury. Thus this first stage of the AWARE study during the past four years has provided us with a lot of very interesting data, which we will use to establish our research network in the next four years and beyond.

CHAPTER 11

What Does It All Mean?

O NE OF THE MOST prominent and highly significant, yet almost completely unknown, articles to appear in the medical literature in the twentieth century was "Le Coma Dépassé." Authored by Maurice Goulon and Pierre Mollaret, this article was published in the summer of 1959 in the medical journal *Revue Neurologique* and described the observations made by two French doctors working at a hospital on the outskirts of Paris. Goulon, the younger of the two and only forty years old at the time, was a veteran of the Second World War, where he had served with distinction as a twenty-year-old medical aide at a hospital located in the quaint town of Vendome, 110 miles south of Paris. After surviving the 1940 Luftwaffe bombing of an airfield adjacent to his hospital, which caused many deaths and gruesome casualties, he took up medicine as a profession. Twenty years later, he had been working for a number of years with Pierre Mollaret, an accomplished physician, to treat patients with neurological disorders who were suffering from respiratory failure, the dreaded complication of paralyzing neuromuscular disorders such as myasthenia gravis and polio.

Under normal circumstances, this gradual weakening and paralysis of the muscles of respiration would invariably lead to death by slow and gradual suffocation. However, by the mid-1950s, physicians such as Goulon and Mollaret had much more hope in terms of saving people's lives, as methods had been discovered that could help paralyzed patients artificially breathe. At their state-of-the-art Centre de Reanimation Respiratoire de L'Hopital Claude Bernard in Paris, which Mollaret and Goulon had helped establish in 1954, they installed rows and rows of large human-sized breathing chambers, called iron lungs, in which patients unable to breathe for themselves were being kept alive. At the time, this particular unit and many others like it owed their existence to the overwhelming need that had been thrust upon the public by the large-scale seasonal polio epidemics that, though largely unheard of before the twentieth century, had devastated the world almost every year since the turn of the century. These outbreaks had caused widespread panic as well as thousands of deaths in major cities, including New York, Boston, Copenhagen, and London. However, while these "reanimation" units, which would later evolve into today's intensive care units, would go on to save countless lives, it was the observations made on this particular medical unit, followed by many others around the world, that would shake our understanding and definition of death and its relationship with the human mind, psyche, and consciousness. But to really understand the implications of these observations and the issue of *coma dépassé* itself, we must look back at the evolution of events that had started some fifty years earlier in response to the rising death toll from an old foe that had resurfaced: the polio virus.

Confronted with the perennial problem of the inability to breathe, which occurred as a complication of polio and many other medical conditions and thus regularly took countless people's lives

every year, after the turn of the twentieth century, a group of researchers took some of the knowledge gained from the Industrial
Revolution and started to work seriously on developing different
forms of artificial ventilation systems as a way to provide breaths to
people when their lungs were not working. Although similar work
had been tried in the past, it hadn't been developed in a systematic
manner until then. In 1908, a physician in Brooklyn, New York,
asphyxiated a dog and then revived it by breathing into its mouth.
The story made the front page of the *New York Times* and led to a
reemergence of ventilation work that had originally started with the
use of a fireplace bellows centuries earlier in Europe. By 1911, the
first portable artificial ventilator, called the Pulmotor, was developed by Heinrich Dräger, and it was carried by police and fire units.
It was a mask that could be placed over a person's face and would
blow air into the lungs. By the 1930s, work had progressed a lot,
and the iron lung that Goulon and Mollaret were using in Paris was
produced. This would enable a person to lie flat in a long iron box—
almost like a coffin, except that the head and legs would stick out—
while the machine would pneumatically suck the person's chest wall
in and out, thus creating breathing movements. Iron lungs were
used a lot in the following two decades, particularly during polio
epidemics that took place all over the world in the 1940s and 1950s.
Polio victims could not breathe because they had been paralyzed
by their illness, and they became dependent on iron lungs to stay
alive. These techniques were refined further when the modern ver-
sion of the ventilator was adopted in the 1960s. Unlike when they
used the cumbersome and very large iron lungs, doctors learned
to insert a breathing tube into a person's mouth and windpipe and
then directly push air into the lungs from a machine. Although this
discovery led to the birth of the modern ventilator and the new field
of intensive care medicine, it also led to a huge ethical dilemma regarding understanding when death becomes final.

The first sign that the definition of death would perhaps need to be reconsidered came around this time. People with massive brain injuries who would normally have stopped breathing were now being kept alive artificially by ventilators. Normally after massive brain injury, the brain, including the part that controls breathing (the brain stem), stops working and so the lungs stop taking in breaths. Then the heart, which is deprived of oxygen, stops beating altogether, causing the person to die. Before the discovery of the ventilator, things were simple. There had been no dilemma, but now the machine would provide breaths even in the context of massive brain injury and so people didn't automatically die anymore. They could be kept alive artificially in the sense that they would continue to get breaths and hence oxygen, and their heart would also continue to beat.

Doctors then started to notice a new and bizarre phenomenon in these massive brain injury patients who were being kept alive on ventilators. Soon after being placed on a ventilator, they started urinating uncontrollably and needed gallons of replacement fluids given by drips every day to keep up (it was later discovered that the brain was no longer producing a hormone that normally regulates urine production). These patients had no brain reflexes and remained in complete comas with no neurological recovery. Interestingly, they all eventually died several days or weeks later when their hearts finally stopped. When their bodies were sent for autopsies, pathologists found that their brains were basically liquefied. They had clearly been alive but had become brain dead and had even progressed to a point beyond, such that their brain tissue had degraded. It was clear from the evidence of the autopsy findings that this had taken place sometime after being attached to the ventilator and after their hearts finally stopped.

This was a new phenomenon in that people could remain alive (because the heart and lungs were working) for some time, despite the fact that the brain had completely died and after many days and weeks had even turned into a gel such that its cells no longer even maintained the structure of brain cells. Thus, there was no possible return at that stage. This was essentially irreversible brain death and had not been seen before, since before then once the brain had stopped functioning (even if it was still in a reversible stage along its own trajectory of death), the person would die when breathing ceased and the heart stopped. Now, even though the brain could stop functioning entirely and even completely die hours to days later as well as go on to partially liquefy weeks later, the heart could still be beating. So when would such a person connected to the ventilator for weeks really have died? Was it when the heart eventually stopped weeks later, or was it somewhere much earlier when the brain had died while the person nevertheless remained alive with a beating heart on the ventilator? People had not been aware of the distinction between reversible and irreversible brain death until this time, but these events were beginning to indicate such a possibility.

This was the observation made in 1959 by the two French doctors Mollaret and Goulon, and they named it *coma dépassé*—meaning a coma that went beyond the deepest coma and one that was irreversible. Mollaret and Goulon then questioned when people really died and asked the critical ethical question of when resuscitation, including life support measures on ventilators, should be stopped. Do doctors classify individuals as alive only if their brains are still alive, or are people alive as long as their hearts are beating? After all, science had indicated that the "seat of the soul" (which had been debated for centuries) actually lay in the brain as Plato had argued and not in the heart as Aristotle had claimed. If so, when do people really die? Are humans simply the sum of the physiological processes that make up the body? Can we therefore define someone dead once these physiological processes have stopped functioning, which is what happens when someone dies in the traditional sense of the word, or do we need to wait until irretrievable and irreversible organ and cell damage has occurred? In which case, when does that really happen? And what of human consciousness, that entity the

Greeks called the psyche or soul? What happens to this during the process of brain death that caused Goulon to ask the ethical question, "Where does the patient's soul dwell?"

FOR MILLENNIA, DEFINING DEATH was easy and straightforward. Nobody needed to worry about what was life and what was death. It was quite clear and obvious: when a person's heart stopped, he or she was dead. It was known that people would die for two main reasons—either their hearts stopped or they stopped breathing. Either way, whichever stopped first (the heart or the lungs) would cause the other organ to also stop quickly, and then the brain would also stop working almost immediately afterward; hence, we could say someone was dead. In essence, when all bodily functions ceased, an individual was dead. Scientists were not aware of a period of time after death in which the organs and cells in the body remained viable and had not yet become irreparably damaged, and hence death could be reversed. We were also not aware of the fairly long time that existed between these two states. The other way that death could happen was if someone had severe trauma to the brain; in that case, the brain would swell up and then start to press on the brain stem, which is where the reflexes are located that regulate the heartbeat and breathing. If the brain stem is compressed, then all the nerves there stop working and the person immediately stops breathing and the heart also immediately stops beating. This is why being shot in the head would cause someone to die. In a sense, the final step was when the heart stopped. If someone's lungs were not working, the person's heart would also stop, or if his or her brain was severely traumatized, then the heart would also stop. Cessation of activity in any of these three organs would quickly lead to death of the person since the function of the three organs is closely linked and stopping one would quickly lead the others to follow.

But now the advances in medicine indicated that for the first time in history, death had to be defined in some other completely different way. This would enable the definition of death to also include the point in time when there is irreversible brain damage irrespective of whether the heart is still beating—hence, brain death—in order to accommodate the growing number of people who could now be kept alive artificially (by maintaining their heartbeats and breathing) even after they had developed permanent brain death.

As we have seen, there is first reversible death, which is when the heart stops beating and the brain stops functioning, after which the brain and every other organ in the body undergo their own processes of cell death, culminating in irreversible death. But brain death is the finale. If someone is truly and irreversibly brain dead, we cannot consider the person alive, even if the heart is beating, because it is in the brain that the mind, consciousness, psyche, or soul—the self—exists. In a sense, the seat of the soul is in the brain. Deep ethical and moral questions about when death, and in particular brain death, had become absolute and irreversible were being raised and discussed by doctors, because it was not known exactly when during the "gray zone" of reversible and irreversible brain death we could consider someone truly dead.

In 1968, an ad hoc group from Harvard Medical School looked at these difficult issues of death. They contemplated the two critical issues that had come to the forefront of the discussion, which were deeply and intricately related. First, when do you declare someone dead? Second, what is the impact on organ transplantation? Clearly, doctors had to be able to define someone as dead in the absolute sense if they were to then use his or her organs for transplantation purposes and hence give the gift of new life to others. There was a huge ethical dilemma with respect to maintaining the rights of an individual and ensuring in good conscience that there would be no way back for that person after death, while also realizing that waiting a long time after death would not only potentially jeopardize the ability to use the person's organs for transplantation and hence give new life to others but also cause great distress to families. From an ethical and moral standpoint, two competing interests and rights had to be respected. First were the rights of the patient lying in a hospital bed with brain damage who was being kept alive on a ventilator, and second were the rights of those people in the wider community who would benefit from transplantation of the person's organs in order to stay alive. These issues, which crossed medical, philosophical, religious, and ethical lines, needed to be balanced. This led to the Harvard group proposing a series of tests to determine when in fact a dying patient's brain was not functioning and had likely become irreversibly dead as opposed to not functioning but in a reversible state.

The tests, called the Harvard Criteria, were published in the *Journal of the American Medical Association* in 1968. These tests looked for unresponsiveness, no movements or breathing, no reflexes, and a flat EEG (no electrical activity in the brain). But doctors would only be able to ascertain if the brain was functioning or not rather than if it was irreversibly dead (tests can only be done to determine brain function and not actual irreversible death, and no test could actually distinguish between the two states of reversible and irreversible brain death). As we have discussed, the brain can stop functioning, but this does not necessarily imply it is irreversibly dead. The brain stops functioning under different circumstances. In fact, anything that impacts the ability of brain cells to be active will cause the brain to stop functioning. This includes when someone's blood sugar has dropped to very low levels, or if the temperature in the body is very low. Certain drugs, particularly those given for sedation and anesthesia, will also stop brain activity if given in high enough doses. So determining when the brain was not functioning due to being irreversibly damaged and dead from a reversible state would be the challenge. The Harvard group concluded that doctors should eliminate all other possibilities for a nonfunctioning brain. The tests also needed to be performed over a period of time to distinguish between a nonfunctioning state, due to being irreversibly dead, as opposed to one that may come back again and is hence reversible. So it was mandated that, before patients were pronounced truly brain dead, the tests first needed to show no function in the brain and then needed to be repeated after twenty-four hours to document that no change had occurred in the result (and hence no reversibility). Doctors also had to exclude patients on drugs that could depress brain function and patients suffering from hypothermia (each of which could mimic what appeared to be brain death) during this twenty-four-hour period.

In the ensuing years, many countries in the world, particularly those with active organ transplantation programs, started to adopt the criteria for irreversible brain death as being sufficient to diagnose someone as being dead even if there was a heartbeat and the patient was artificially breathing on a ventilator and thus being kept alive. This was in addition to the usual method used to define death, which involved circulatory arrest after the heart stops. Thus people currently accept that there are two different ways to define someone as being dead. The first is the traditional definition of no heartbeat, no respiration, and no brain activity (hence the loss of all function after the heart stops without necessarily having irreversible cell and organ damage), and the second is the more recent definition of irreversible brain death irrespective of whether the heart is beating or not. Interestingly, though (and perhaps somewhat concerning in view of its implications), the criteria used to determine brain death and how frequently the tests should be repeated and by whom (such as how many doctors) continue to vary from country to country and reflect particular cultural beliefs and fears while trying to prevent "getting it wrong" and diagnosing someone with a reversible brain disorder as having irreversible brain death. Therefore, to this day there is no uniform way to define irreversible brain death globally, and one of the biggest differences has been between the United Kingdom and the United States.

In the 1970s in the United Kingdom, a joint commission was

formed by the Royal College of Physicians to examine the issue of brain death. The commission came to the conclusion in 1976 that a person should be classified as dead if the individual's brain stem is dead. If the brain stem (the base of the brain that regulates heartbeat and breathing) is dead, the commission determined that a person couldn't stay alive, irrespective of what is happening in the rest of the brain and whether or not it is alive.

By contrast, according to Dr. Eelco F. Wijdicks, an expert on the determination of brain death from the Mayo Clinic in the United States, "in the USA, many experts felt that brain death could be only determined by demonstrating death of the entire brain." He explains that in the history of further refinement of U.K. and U.S. brain death criteria, one particular period stands out that would bring about an apparent transatlantic divide. On October 13, 1980, the British Broadcasting Corporation aired a program entitled *Transplants: Are the Donors Really Dead?* Several U.S. experts not only disagreed with the U.K. criteria, but also claimed that patients diagnosed with brain death using U.K. criteria could recover. The fallout of this television program was substantial, as indicated by a media frenzy and a six-month period of heated correspondence within the *Lancet* and the *British Medical Journal.* Members of Parliament questioned the potential long-term effect on the public's trust in organ transplantation.

Given the concerns raised, the British Broadcasting Corporation commissioned a second program, which was broadcast on February 19, 1981, entitled *A Question of Life or Death: The Brain Death Debate.* Two panels debated the issues on the accuracy of the electroencephalogram and its place, the absolute need for assessing preconditions before an examination, the problems with recognition of toxins, and the feasibility of doing a new prospective study in the United Kingdom, which would follow patients' examinations assessed with U.K. criteria until cardiac standstill. The positions of the United States and United Kingdom remained diametrically opposed to each other.

These events and the academic and public stir caused by the differences highlight the ethical dilemma and sensitivity faced in determining absolute irreversible death, while also recognizing the need for its timely determination in order to enable organ transplantation in those whose lives will be changed by the gift of life that comes with the donation of a new organ. Although the Royal College responded with a series of articles and research over the years to both demonstrate that the BBC's conclusion had been incorrect and to regain the public trust, the events highlight some of the difficult ethical issues at hand and the fact that differences in opinion exist that reflect even on the most fundamental issue of all: the definition of life and death.

The first comprehensive study to examine when brain death was likely to be permanent was funded and conducted in 1977 by the National Institutes of Health in the United States. It followed 503 patients whose brains were unresponsive and tried to determine the point of irreversible death. The study concluded that six hours of complete brain inactivity in all areas were needed (not the twenty-four recommended by the Harvard Criteria) to declare a person brain dead.

In 1981, a presidential commission published guidelines based on all the research done that provided a concrete basis for declaring a person dead. The commission defined this as "the irreversible cessation of all function of the brain including the brainstem." The periods of testing this activity were set at six hours in cases with no blood flow to the brain or electrical activity, twelve hours for an established and irreversible coma, and twenty-four hours for anoxic brain damage (brain damage due to a lack of oxygen) as occurs after a cardiac arrest. This resulted in the Uniform Determination of Death Act, which stated: "An individual who has sustained either (1) irreversible cessation of circulatory and respiratory functions, or (2) irreversible cessation of all functions of the entire brain, including the brain stem is dead. A determination of death must be made in accordance with accepted medical standards."

Around the world, different countries were coming to different conclusions when they looked at the ethics and morality of declaring someone irreversibly brain dead. While the U.S. guidelines mandated that the entire system regulated by the brain not be functioning, the U.K. position maintained that if the brain stem did not work, then the person was brain dead and hence could be declared dead—in both cases with or without breathing or a heartbeat. The main point was that when someone had developed permanent brain death, then doctors would not have to wait for days or weeks for the heart to naturally stop itself while the patient was on a ventilator before declaring the person dead; they could do so even with a beating heart, and thus they also could remove the person's organs and use them for transplantation in others.

To this day, there are disparities across the world in making the determination when someone has permanent, irreversible brain death. According to Dr. Wijdicks, "The United Kingdom criteria for brainstem death permeate in the previously colonized countries, and Central and South American countries generally follow the United States position on whole brain death." Most European countries have criteria that are similar but have some differences between them. Asian and Middle Eastern countries also have specific criteria while most African countries do not yet have such criteria. The other big differences among the world's countries are concerned with how to determine the diagnosis of brain death (which tests must be performed and how long doctors must wait between tests to conclude that death is not reversible since the tests can only determine function). In some nations, doctors must wait three days before declaring someone brain dead, whereas in others there are no real criteria. Another subcategory is the number of examiners required to make this declaration, and this changes from country to country and in the United States from state to state.

The state of New York, where I practice medicine, changed its law recently. In the latest document published in November 2011, it is stated that "New York State regulation defines brain death as the irreversible loss of all function of the brain, including the brain stem." However, in this edition of its guidelines, the state has removed the requirement to have two doctors perform brain testing and now requires only one doctor. There is also ambiguity regarding the optimal time that doctors should wait before performing brain testing. Whereas previously the state had set a six-hour minimum waiting period before carrying out tests aimed at determining whether loss of brain function had become permanent and hence irreversible brain death had set in, in the 2011 edition it is stated that physicians should carry out the tests after "an appropriate period of time" that is "sufficiently long as is relevant to the individual patient's condition [in practice, usually several hours], after the onset of the brain insult to exclude the possibility of recovery." This new wording simply reflects the fact that different brain conditions require different periods of time but that the absolute correct time for these conditions is not known. The guidelines further state that "adding a waiting period early in the process—before the clinical assessment of brain death is initiated—provides greater assurance that there is little potential for improvement and is consistent with current clinical practice. Only after it is clear that the patient will not likely recover should the brain stem reflex test and apnea test be conducted.

"In all cases, these guidelines advocate a high degree of caution and vigilance to ensure that there is no possibility of a patient's recovery." Thus, while the guidelines acknowledge that there is the possibility of recovery even after patients appear to have reached the criteria of brain death using current tests (since the tests only determine function and not irreversible cell and organ death), they are not specific about how long physicians should wait to be absolutely certain that their patients really can't recover from their brain injury and have thus reached the stage of irreparable and irreversible damage. In practice, then, these guidelines will very likely lead to different criteria at different institutions, and, most important, they don't consider the state of the growing number of patients who have received hypothermia treatment after cardiac arrests, which is known to considerably alter recovery times for the brain. People who have been cooled down after anoxic brain injury from cardiac arrest may demonstrate absolutely no brain function many hours not only after being cooled but even after they have been warmed up to a normal temperature.

Therefore, if someone performs brain testing on such people, say, six, twelve, or even twenty-four hours after they have been warmed up, they may still have no brain function and thus meet the current criteria for brain death but could actually be in the reversible stage of death. This is because the brain takes longer to regain its function after being cold, and no one knows exactly how long to wait before performing tests aimed at determining permanent brain death in this group, except that it is becoming very clear that we need to wait much longer than many of us had anticipated. There are in fact case reports of people who had appeared to be brain dead (and had met all the brain death testing criteria) after being examined many hours and days after being warmed up to a normal temperature following hypothermia treatment for cardiac arrest, only to show signs of brain recovery up to seven days later. Therefore, if physicians are not fully aware of this possibility, they may determine that permanent brain death has in fact set in (based on the results of the brain death testing that had been carried out) much earlier and thus tell families that their loved one is in fact "dead" (since using the revised definition of death, "death" and "brain death" are synonymous even with a beating heart) and they should consent to the withdrawal of life-support measures. According to my own personal experience as well as based on publications and work presented at major conferences, many physicians are not aware of this fact and thus determine that patients have reached permanent brain death much sooner and thus withdraw life-support measures including ventilators while the patients are still within the potential brain recovery times after a cardiac arrest. This is why in my opinion having two people perform the tests at least provides a safeguard against this possibility.

Although the latest New York State guidelines do acknowledge that having received cooling therapies, such as in Joe Tiralosi's and Dr. Kelly Sawyer's cases, will potentially alter the optimal timing of the tests designed to determine brain death, they don't specify how long people should wait. They simply state "where hypothermia was induced previously in a patient, additional vigilance is recommended. In such cases, a prolonged waiting period after the rewarming phase is completed may be appropriate." The lack of more clear and precise recommendations and legislation thus paves the way for personal interpretations and individual forms of practice that most people would agree we should try to avoid, particularly bearing in mind the implications of a mistake on someone's life. Reliance on older data obtained from studies carried out before the widespread use of cooling therapies (such as the time of Mollaret and Goulon or the results of studies that were used for the Uniform Determination of Death Act published in 1981) is also currently highly problematic, because the patients enrolled in those studies did not receive postresuscitation treatment, including hypothermia, that slows down the process of cell death in the manner that is recommended today. All those studies, which are now largely obsolete, would have to be repeated in light of the impact that cooling has on the progression from reversible to irreversible brain and other organ death. What we do know is that, while at the time of physicians such as Mollaret and Goulon the brain had not only definitely died after a few weeks but had even gone beyond and in some cases even liquefied in parts, the answer today to the question of when exactly permanent and irreversible death sets in is still not known. The reality is that the longer we wait before performing brain death testing, the more likely it is that someone whose tests show no evidence of brain function has really reached irreversible brain death and that currently available treatments such as hypothermia as well as other potential future therapies, including Dr. Neumar's drugs that slow down the rate of brain cell death by blocking the chemical catalysts and enzymes involved in this biological process, may radically alter this process and thus extend the gray zone much further.

So WHAT DO ALL these advances in the science of resuscitation mean for us today, and what do they mean for our understanding of life and death? Are the dilemmas we are referring to simply a result of inaccurate human-made definitions? Are we dealing with semantics and just need to redefine death? Is the problem that our definition of death for centuries has simply not been accurate enough?

Clearly if we consider death as the philosophical "end"—a time when nothing else could exist—then perhaps we should frankly just redefine death and move the goalposts to a time when a person has progressed well beyond reversible cell death and reached a point where there is irreversible brain and other vital organ cell damage and death, such that the death of a person has really become irreversible, no matter what we do. To take this to an extreme and just for illustration purposes, obviously nobody would consider an Egyptian mummy to be in any state other than irreversible death, but the problem is that reaching the irreversible stage could take quite a long time.

Although we estimate for brain cells this could be many hours after the heart has stopped and a person has died in the traditional sense of the word, the truth is that from a scientific perspective we may not know exactly what that point is.

But if someone asked me the question "Exactly when after death does irreversible death become an absolute reality, and when exactly does death become completely permanent?" I would answer, "We don't really know, and whatever point we choose will still likely be arbitrary and most likely will need redefining again in the future as science and technology improve and our ability to resuscitate people progresses." Today, we could arbitrarily say that we will define death as the moment after the heart has stopped and the brain has undergone, say, 50 percent irreversible cell damage—perhaps eight or so hours after a person becomes completely lifeless, their heart and respirations have stopped, and the pupils of their eyes have become fixed and dilated and hence their brain has stopped functioning (in essence, how we define someone dead today). But aside from the fact that waiting that many hours after someone has died to define the person as having reached a point of irreversibility is a very long time and will cause great distress to families, what would then happen when we develop systems in the future to resuscitate the brain cells even after 50 percent have become irreversibly damaged? What if the results of work carried out by physicians such as Dr. Robert Neumar do lead to the development of a whole host of new drugs that can be injected into people after they have died that would slow down the rate of enzyme-driven chemical reactions in the brain, such that brain cell death is slowed down greatly, thus extending the gray zone between reversible and irreversible death? In short, what if what we consider irreversible today becomes redefined as being reversible tomorrow? How do we then define death? Do we just keep moving the goalposts to another physiological point that may then be conquered by others in the future?

Nothing brought this point home to me more than when I recently watched again James Cameron's famous blockbuster movie *Titanic*. I keyed on the scene toward the end of the movie that showed the panic and chaos on board the ship after it struck the iceberg and was sinking. Although I had first seen the movie fifteen years earlier, something instinctively struck me this time when I watched those final scenes again. Historical records show that after hitting an iceberg the RMS *Titanic* sunk at approximately 2:20 A.M. on April 15, 1912, with the loss of 1,514 lives. There hadn't been a sufficient number of lifeboats on board, and as a result, many people were thrown into the icy waters of the Atlantic (which were around –2°C [28°F]) after the ship had sunk. The first rescue ship to arrive at the scene was the RMS *Carpathia* approximately two hours later at 4:00 A.M. In the movie scene, the crew of the *Carpathia* were desperately looking for survivors from the water, only to find hundreds of dead bodies floating in the icy cold water. Thus, the news reports at the time correctly reported the tragic loss of 1,514 lives. However, now (even between 1997, when Cameron's movie debuted, and 2012) a major change has taken place in our understanding of death and resuscitation, which is that if people are cold after death, it can preserve their brain cells, and as a result hypothermia is now commonly advocated as a form of treatment for this purpose. Furthermore, we know the transition from reversible to irreversible death also depends on many other factors, but if someone's body temperature drops down after death, then the speed by which the cells reach the point of irreversible damage and death slows down considerably.

So what really struck me as an intensive care physician while watching the film again was that today we would not have necessarily declared those people dead—at least not in the irreversible and irretrievable sense. Although I agree they were dead, they were nonetheless salvageable. Their bodies would have been largely preserved by the icy cold waters, and two hours is not much time at all. In short, they were potentially completely viable. In fact, their bodies were probably somewhere near the ideal temperature that would have preserved their cells really well, since body temperature normally drops by about one or two degrees per hour after death. Today, people who die in a cold climate or those who die in a cold lake can be revived even many hours after dying, as illustrated by the case of the person who was found dead in a forest in Japan after an overdose. She had been dead for so long that her body temperature had gradually dropped from around 37°C (98.6°F) to around 20°C (68°F) over many hours in response to the outside temperature, until she was found dead by ambulance staff at around 8:30 in the morning. Yet she was successfully revived and eventually left the hospital with almost no brain damage more than twenty days later.

What a difference a hundred years makes, and what a different headline we could have expected if the *Titanic* had sunk in April 2012, instead of April 1912, especially if perhaps instead of the *Carpathia,* a rescue ship with a team of people trained in the nuances of resuscitation had come to the rescue. But sadly as we all know, that isn't what happened, and watching all those lifeless bodies floating in the icy cold waters at the end of the movie, I couldn't help but think, *So when exactly did those unfortunate people die?*

WE SIMPLY DON'T HAVE all the answers. But we do know that the once-held philosophical idea that there is no way back after death is not accurate and that there is a significant period of time after death in which death is fully reversible. The goalposts have moved, and we don't know where the science will take us. Perhaps we should actually define death not by some arbitrary physical point or moment linked to a specific physiological change in the configuration of certain cells whether in the brain or otherwise, but really in terms of when a person's consciousness, mind, psyche, and soul—the self— has been lost forever and cannot be retrieved again. After all, we are all conscious, thinking beings, and it is this that defines who we are. This is a deeply important issue that many have grappled with for years, and it is this question that we have been contending with for more than half a century since Goulon's time. As an intensive care physician dealing with these issues on a daily basis, I also often find myself under the weight of the same moral and ethical dilemmas that Goulon faced, which must have ultimately led him to ask the question, Where does the patient's soul dwell in all this?

Perhaps today we would agree that, in view of the rapidly evolving progress in the field of resuscitation science and the ever-expanding gray-zone period after death, it is important to include what we would refer to as human consciousness, psyche, or soul in future definitions and considerations regarding death. It would also perhaps be wise to concentrate some of our future research efforts on understanding the state of human consciousness after death has started, since the evidence currently suggests that it is not lost immediately after death but continues to exist for at least some time afterward.* This is what the case of the man with the profound out-of-body experience during the AWARE study also seemed to suggest. I appreciate that this may be a challenge to some people's perceptions of what science should be, but after studying this field for over fifteen years, I am convinced that this is not a challenge we should be afraid of. The real challenge probably lies in realizing that while we humans naturally have to establish limits and hence define what we consider science (as opposed to what we define as not being science) based on our own abilities and limitations at any given era, science, the objective method of learning about the entirety of the forces and realities that exist within the universe (whether within the reach of human perception at any given era or not), is itself limitless. It will thus routinely challenge our inherent limitations and in the process help us unearth unexpected discoveries that we could never have anticipated—that is, as long as we are willing to look beyond the limitations we may have set for ourselves. The challenge for a scientist is thus to pursue what may at first seem implausible or unlikely in the hope of finding an incredible piece of treasure in the form of greater knowledge and understanding. Think of the pursuit of the dream of flying, space travel, and even the battle against age-old killer diseases (such as polio), to name just a few. None of these realities would have been accomplished without many people fol-

---

*As opposed to disappearing transiently and coming back after the modulation of the relevant brain circuits by conditions such as the administration of a general anesthetic or medical conditions that lead to a lack of oxygen to the brain (which is also what occurs when we first die) as discussed in chapter 9.

lowing leads that to others must have seemed unorthodox, wishful thinking, or just impossible dreams at the time. However, we must also realize that during this pursuit of knowledge, many of the leads we follow will most likely turn out to be of little significance. Nevertheless, as with those who deal with the prospecting and mining industry, scientists have to follow various leads because all it takes is to have one lead turn into a gem for it to have all been worthwhile, as happened in South Africa just over a century ago.

ALONG THE NORTH BANK of the river Thames in central London lies a historic castle made up of a complex of several buildings set within two concentric rings of defensive walls with an exterior moat, known as the Tower of London. Dating back to 1066, this castle was used for much of its existence as a prison, most notoriously as the place where, among others, Elizabeth I was incarcerated before she eventually became Queen of England. Throughout the years, the castle has also been a royal residence, as well as home to the royal mint. Today, it houses the British monarchy's crown jewels, including the famous Cullinan I diamond, a 530-carat gem also known as the Great Star of Africa, which for much of the twentieth century was the largest polished diamond in the world. What many people who visit the Tower of London every year may not be aware of is that had it not been for a very observant miner and a major risk taken by a construction worker, this incredible diamond would have possibly never been discovered.

After gaining wealth, Thomas Cullinan, a relatively unknown South African bricklayer and later building contractor who was aware of the newfound riches that had been afforded to Cecil Rhodes, founder of the De Beers diamond mining company in 1888, decided to also try his hand at prospecting. He took a risk by paying fifty-two thousand pounds for a piece of land near Pretoria that many others believed had no real potential since it was away from the usual diamond-mining areas that others, including the De

Beers group, had discovered. Nevertheless, despite other people's opinions, Cullinan had faith in the land he had purchased, and he went on to establish his own rival mining company in 1902. Just over two years later, in January 1905, Frederick Wells, the superintendent of the mine, was making a routine inspection late one afternoon and noticed that the reflection of the sun coming off one of the walls of the mine was a little unusual.* It was as if something in the wall was causing the unusual reflection. He climbed up the wall and managed to dig around the area where the reflection was coming from and eventually extracted what appeared to be a very large crystal. Initially, he thought it was perhaps just a large piece of glass or maybe a worthless piece of crystal, and so out of fear of ridicule, he didn't even tell others about it. He did, however, send it for testing just in case it was something worthwhile. Later, the report came back that at 3,106 carats, this crystal was in fact the largest gem-quality diamond ever discovered. The rough crystal was named the Cullinan diamond after the mine owner and was sold to the government for 150,000 pounds, which then presented it to King Edward VII as a gift in November 1907. Being the size of a man's fist, the diamond was eventually cut into nine pieces, and the Cullinan I, which lies in the Tower of London today, is the largest of these nine pieces.

SCIENTIFIC EXPLORATION IS IN many ways not dissimilar to prospecting and mining. While at any moment in time, certain scientists have pioneered and paved the way forward and have thus established a legacy that others then follow, many others continue to explore new avenues that may at first seem either of little worth or even completely unorthodox and, in short, are considered an impossible dream in the eyes of other people. Like prospecting, it would prob-

---

Some reports state that the diamond was actually discovered by another miner, Thomas Evan Powell, who then gave it to his superintendent, Frederick Wells.

ably be true to say that much of what they explore may end up having little or even no return, but without exploration of these completely uncharted new areas and taking chances as Cullinan had done, many major discoveries will not be made. Cullinan took a chance where others did not.

Thus, whereas many scientists continue to "mine" in the same areas that have already been discovered, many others "prospect" in different areas in search of new discoveries. For instance, many people don't realize that Sir Isaac Newton, the father of modern physics and a man who is universally recognized as one of the greatest scientists of all time, actually spent much of his time exploring a whole manner of unorthodox practices, of which alchemy was perhaps the most notable. In fact, according to some historians, Newton, like Flamel before him, was first and foremost an alchemist, and much of what he studied was so unorthodox that it would possibly be classified as "occult" today. He presumably believed that what he was pursuing would be worthwhile, but what is certain is that many of those things did not actually lead to anything long lasting. However, if it hadn't been for those explorations, he would not have discovered the scientific treasures for which he is famous today since the universally applicable laws of physics that he described in relation to optics and motion came about as a result of his overall pursuits.

TODAY, THE QUESTION OF consciousness, psyche, and soul is a completely new area of discovery that, although an enigma, has thankfully become a point of major focus and interest in science. To better explain the scientific situation we find ourselves in, it is as if we have discovered a wholly new type of substance that we can neither account for nor even explain in terms of anything we have ever seen and dealt with before in science. It is not like understanding the science of cell function, or for that matter any other entity we have studied in the physical sciences in the past. Though it is a pure mystery, we know it exists and defines who we are; however, as explained earlier, nobody has been able to explain how human consciousness comes to be. Broadly speaking, there are essentially two ways to account for its existence. Consciousness owes its existence to either a "down up" phenomenon, which essentially means our consciousness, psyche, or soul is a by-product of brain cell activity—an epiphenomenon—arising from the coordinated activities of certain brain segments, or a "top down" approach that considers consciousness, psyche, and the soul to be a separate entity that, while undiscovered by science today, is not produced by brain cells and can itself independently modulate brain activity. Ultimately, whichever model one chooses, the human brain and that which we consider consciousness, psyche, or soul are intricately involved with one another, and as such "the seat of the soul" does indeed lie in the brain. But what still remains to be answered is the question of the nature, origin, and relationship of consciousness with the brain. This is why identifying an association between brain cell activity and the mind using modern brain-scanning techniques only provides us with a correlation regarding their close relationship but not evidence that consciousness or the soul is produced from the brain.

As a consequence of the progress in resuscitation science that started in earnest almost half a century ago and has been evolving since, over the past forty years there has been growing recognition that people who have had a close brush with death or have gone beyond the threshold of death and entered the gray zone that exists between death and permanent irreversible brain damage have provided consistent mental recollections that correspond with that period. Today there have been countless accounts described from all over the world by people from different backgrounds, cultures, societies, religious convictions, and age groups. It is true that often people have interpreted their experiences (as with any other experience) based on their own preformed psychological concepts of reality (e.g., one person may interpret seeing a being of light during an ADE as Jesus or God, while another person defines it as Krishna and another doesn't give the being any name). People in general "see" and interpret things through the prism of their own minds. Furthermore, it is also true that the impact of a lack of oxygen on memory circuits in the brain means that people's recollections of their experiences are often fragmented such that most people usually only recall some of the features associated with a complete ADE or NDE. As we have seen with the AWARE study, the only people who have recalled out-of-body experiences so far have had short periods of cardiac arrest, suggesting that the postresuscitation tsunami effect has been less profound. As a result, their memory circuits may have been less changed and better preserved, thus enabling them to recall the experiences better. Nevertheless, the one unifying factor in all these experiences is that they are remarkably consistent and universal.

Interestingly, what complicates our ability to explain these experiences using current scientific models is that, in general, after people have entered the gray zone beyond death, brain function ceases almost immediately as the cells start to undergo their own process of death (even if they have not yet reached that absolute final point of irreversibility). Thus, these experiences appear to be somewhat of a scientific paradox, and they raise questions regarding how people could have lucid, well-structured thought processes with memories and recollections during this period without a functioning brain. In short, if one adopts the bottom-up model of consciousness, psyche, and soul, then cognitive and mental events should not be possible at this time without a functioning brain, unless what we are dealing with is either a top-down model of consciousness or some other undiscovered process or an error of timing related to the experiences. Furthermore, another question arises: What does the suggestion that memories may be formed at a time when there is no function in the brain tell us about the nature of memory and its relationship with the brain? That is, what does it tell us about memory and the role of electrical impulses that normally arise from the brain and are considered the hallmark of communication within different brain regions? Does the brain resemble the hard drive on a computer, which actually stores memories in itself, or is it more like the RAM, which, though required to perform tasks and functions, does not store memories in and of itself? If the brain is the hard drive for our memories, how then can memories be generated and stored when the brain is not functioning and there is no electrical activity? On the other hand, if the brain is more like the RAM on a computer, does that mean that memories can be stored in our consciousness, psyche, or soul even in the absence of brain function, and is this maybe what is happening during an actual-death experience?

These fascinating questions have arisen as a result of the observations in the AWARE study that my colleagues and I have made so far. Even though we don't have all the answers, at the very least these issues warrant a serious scientific exploration and an open review of our models and concepts regarding human consciousness, psyche, or soul, and the brain. When we started the AWARE study, many people representing both extremes of the spectrum of opinion were intrinsically against the idea of any type of study of near-death/actual-death experiences or consciousness during cardiac arrest or anything else of the sort. Many felt they already knew that it was impossible for consciousness to continue beyond the threshold of death and that any question to the contrary would be absolute nonsense. They perhaps felt that we were trying to perform a "spiritual" study that was therefore by definition in their minds completely unscientific. They further believed that there was no need for any so-called supernatural explanations for people's experiences, since they could all be completely explained as tricks of the brain. At the other extreme of the spectrum of opinion, there were many people who were already convinced of the reality of so-called paranormal and parapsychological events, and they considered NDEs and anything else that may resemble them to firmly lie in the realm of their subject matter. For instance, some people were convinced that in view of the numerous challenges that our study would face, we had probably set it up for the specific purpose of discrediting people's reports and wanted to ensure the study would scientifically "fail" so that we could "prove" that the experiences people recall are unreal and simply hallucinations.

The reality, of course, is that we are merely trying to remain objective as scientists and are seeking to understand the truth while trying to work within the inherent limitations that exist for all of us. To go back to the previous analogy, we realized that although many scientists have focused their efforts on seeking answers by "mining" in previously explored areas of brain science, which we agree is important (and we too have also worked on other conventional areas as part of the AWARE study), we had taken a chance by also "prospecting" in new areas that may seem unusual to some other people, knowing full well that the results may actually not lead to anything long lasting. We simply believed when we started that it is important to explore this area, and today we are at the stage where Frederick Wells found himself after he noticed an odd reflection coming off the wall of the mine. Clearly, statistically speaking, as he started to climb up the wall, the most likely outcome of his closer examination would have been that the reflection was probably nothing more than an unusual rock or something similar without too much value. Today, we too don't know exactly what the reflection is that is coming off the wall of cardiac arrest studies, but unlike many others, we just believe it is worth exploring. Through the results of the AWARE study so far, we realize we are only at the stage of starting to dig around the area of the wall. We have not yet been able to fully dig around the area and thus don't know exactly what we are dealing with. However, we think it is important to complete the search and explore the area irrespective of whether or not what we have observed turns out to be something of a gem.

NOW WE UNDERSTAND THAT, contrary to what we used to believe, it takes at least many hours, and possibly even longer, for irreversible brain death to set in after a person dies, which gives us a much longer window of opportunity not only to pull people back to life after they have died but also to study what people experience during the first period of death. Today, these survivors have gone on to provide us with remarkable yet consistent accounts of what they experienced during the period after their own deaths had started. Now, the problem for us, as scientists adhering to the principle of objectivity and while remaining unbiased, is, What are we to do with the accumulating evidence provided to us by these millions of survivors from all over the world? How do we accommodate these experiences, which in essence represent revealed insights into our objective scientific models?

It is true that throughout history, the discussion of what happens when we die had in general been limited to only that which could be perceived through revelation, since the scientific means that would enable humans to explore this subject in an objective scientific manner had not been available until now. This is probably why there had always been so much disagreement and subjectivity associated with any discussion around the subject and also why some rational-minded people probably dismissed the whole area as being nonscientific. The challenge for science today is trying to understand how to reconcile knowledge that may have come about through revelation (with all its inherent limitations) with knowledge garnered from the new modern experimental scientific methods that we have at our disposal, such as functional magnetic resonance imaging (MRI) of the brain. Should we simply reject any and all forms of understanding that have arisen from revelation? If so, what are we to do with the millions of near-death and actual-death accounts? Should they be excluded from any analysis of what happens when we die?

I believe that when examining what happens when we die, most objectively minded people would consider information that derives from two main sources. The first is revelation, which, as we have seen, has been provided to us by many millions of people from all over the world, and the second is conventional experimental scientific methods using the tools that are available to us. The challenge is thus how to bridge these two sources. Although it may be easy to dismiss the accounts provided by revelation, the problem for an objectively minded scientist today is that a sufficient number of compelling anecdotes have been reported to at least warrant a serious scientific look at the issue. One of the challenges, of course, is that there are rarely scientific methods or tools available to verify the reality of any human experience (whether related to death or otherwise). Of course, the alternative, which is to simply dismiss the millions of people's reports of experiences related to death due to the lack of such tools, is also unscientific. Thus, as scientists we cannot readily ignore or dismiss these accounts, since they provide a valuable resource in terms of understanding the mental and cognitive aspects of what happens when we die; but we do need better tools to study the neurological elements that underlie consciousness and human experience. In support of the need for absolute serious scientific research, it is perhaps also important to point out that even though dogma is typically associated with religious belief, an unwillingness to acknowledge new and emerging evidence in scientific fields may also be construed as dogma, which if not resolved could have the same detrimental effects on society that other strongly held dogmatic beliefs have had in the past. This is perhaps why a number of researchers and physicians, from many different backgrounds and stripes working on the front line of patient care, have agreed that the mental and cognitive phenomenon that follows after people have died and the questions it raises should be studied in a serious scientific manner.

I have spent the past fifteen years extensively studying the objective and experimental science pertaining to what happens when we die as well as the closely related subject of the human psyche, consciousness, or soul. Parallel to my scientific research, I have also comprehensively studied the broader areas of knowledge that relate to these fields, including the philosophical literature. Understanding where we have come from helps us better understand our own beliefs today as well as current scientific opinions and where we are going. This is why I have outlined some of the major schools of thought relating to this subject in this book. An interesting point that perhaps illustrates some of the issues at hand when trying to bridge knowledge revealed throughout history with our new scientific tools is that many ancient traditions and civilizations, such as the Egyptians and in particular the Greeks, have of course also tackled issues that have been discussed in this book. Even though much of what they discussed may not have been proved from a scientific perspective, there are actually many issues that scientists have come to agree on today. In fact, much of our modern scientific method rests on the observations and systems developed by the ancient Greeks. Thus, one has to ask how it would have been possible for some individuals to come to certain conclusions and understanding regarding issues by some form of personal insight and revelation that are then proved to be correct through the results of objective scientific exploration many years later (a good example is the concept of the atom that was proposed by the Greek philosopher Democritus approximately twenty-four hundred years before it was discovered by modern scientists).

Interestingly, although throughout the millennia many philosophers and scholars have expressed an opinion regarding what happens when we die, the most complete account that corresponds almost exactly with today's scientific discoveries has come from the late Ostad Elahi, a contemporary philosopher and jurist whose work addresses the lives of modern-day human beings and their existential concerns and aspirations.* In fact, for those interested in a spiritual or metaphysical perspective on the survival of our consciousness beyond death, one of the most inspiring and revealing viewpoints I have encountered is that of Ostad Elahi. What astonished me was that long before our scientific discoveries led to the understanding of what happens to the human body when we die and the importance of distinguishing between reversible and irreversible death (which many physicians have still not grasped today), Ostad Elahi had stated the following:

> *When a person first dies, he is not yet dead; it is the heart that has stopped functioning. Although his physical faculties have died, the individual organs in his body (such as muscle and skin) each have their own specific powers to keep the organism alive. These powers can remain alive for up to three days, though they can also perish sooner.***

In these few lines, this philosopher and jurist with no medical or scientific training appears to have summarized the essence of the fundamental issues we have come to discover in medical science today.

THE REALIZATION THAT THERE is a fairly extensive period between the transition from reversible death to irreversible death takes us back to the importance of establishing the most appropriate care for the countless nameless individuals whose lives and brains will be placed in our hands in the coming days, months, and years. Furthermore, one day each of us will have a cardiac arrest; this is for certain. So how would we like to be treated ourselves? We are

---

*For more on the life and work of Ostad Elahi (1895–1974), see Ostad Elahi, Knowing the Spirit, trans. James Winston Morris (New York: State University of New York Press, 2007).
**Words of Truth: Sayings by Noor Ali Elahi, vol. 2 (Jeyhoun, 1991), p. 395.

not just talking about something that will impact our neighbor next door, but ourselves too. Our new understanding regarding the reversibility of death also provides us with a unique window of opportunity such that interventions that we make or fail to make will impact our ability to save lives and brains. It is hard to know that in many instances things that could be done are not being done due in large part to a systems failure in most if not all countries, whether rich or poor. However, these interventions, although often seemingly fairly simple, are quite complex and require specialists with expertise in resuscitation and specifically all the latest developments in the field of cardiac arrest. Unfortunately, though, such specialists are not routinely trained under the current medical systems that exist. True success will only come if diligent attention is paid to every detail of the process of resuscitation—for it is in the details that we will have success—as in the aviation industry.

Even one missing link in the chain of survival can and often is fatal. We need to transition away from a system of care in which any doctor in any hospital can care for the person who has undergone cardiac arrest and death (and is by definition the most critically ill patient in any hospital) to a system in which specialists with absolute expertise (guided by external regulation) regarding all the minutiae of care needed to prevent irreversible brain and other organ damage are at hand to deal with the very complex medical condition that arises after the heart has stopped and with the postresuscitation disease that ensues, which can devastate the brain and other organs without proper attention. It is time to develop optimized systems of care to ensure that everyone receives the best care through a system implemented and monitored similar to the aviation industry, so that we can achieve rates of success that mirror what has been achieved in this industry. If we don't do this, we will fail in our duty to protect people who have had cardiac arrests but could be revived successfully without brain damage. Based on my experience, I believe this can only be achieved if medical funding bodies such as Medicare in the United States or the National Health Service in the United Kingdom provide financial incentives to institutions that provide optimal care, while providing financial penalties to institutions that do not achieve optimal care. Unfortunately in medicine, as with many other industries, financial incentives rather than moral incentives alone play a great role in ensuring that the highest standards are followed.

It has been estimated that if we combine all cardiac arrest resuscitation attempts in the United States alone (between in hospital and out of hospital), there may be around one million events that take place per year. Although of course for many people this could be the natural progression to death that cannot be reversed, for many others much more can be done, as illustrated by the 2009 report compiled by the American Heart Association. That report, as explained previously, concluded: "Organization of the system of care appears to have a larger effect on survival than patient factors. The creation and maintenance of an effective system for delivering optimal emergency medical care are complex. Examining either systems with historically good outcomes or systems in which change has improved outcomes provides an opportunity to identify best practices that can be broadly implemented." In other words, the problem is not that the patients who have cardiac arrests are very sick and probably won't make it. If there is a system in place, survival rates will increase. Even if we can increase survival by just 5 percent across the board, potentially fifty thousand more lives will be saved per year. If we then extrapolate the numbers to other countries around the world, it is clear that millions of lives and brains could be saved. It is important to remember that each person saved is a human being with his or her own story, much like the stories of Dr. Kelly Sawyer or Joe Tiralosi, and that each life lost will affect many others. Many breadwinners could go back home, and fewer children would lose a parent. This is the unrecognized silent problem that affects every nation every day.

FINALLY, WHAT OF THE fear of death and the state of consciousness (or soul) after death has set in? At the very least, today we realize the experience of death does not seem to be unpleasant for the vast majority of people.

That entity that we define as consciousness, the soul, or the self—that which makes me who I am—does not stop existing just because someone has entered the period beyond death. Furthermore, those who have entered this period and been revived and brought back to life (before developing irreversible brain damage) can thus share their experiences with us. It appears that they have taught us that what we experience at death is universal and has bearing for us all. We are likely to have similar experiences after death has started, irrespective of race, creed, or culture; and our real self doesn't seem to be lost in the early part of death.

Finally, if the mind—consciousness (or soul)—can continue to exist and function when the brain does not function after death, then it raises the possibility that it may be a separate undiscovered scientific entity that is not produced through the brain's usual electrical or chemical processes, as we understand them based on today's understanding of neuroscience. If further confirmed, this calls for a new paradigm in neuroscience to address the issue. It requires an objective and scientific approach to understanding the nature of human consciousness or soul, including an objective scientific method leading to the correct and balanced development and growth of the moral and ethical dimension of consciousness that exists in all human beings. Such an approach will no doubt bear fruit for the entire world, leading to greater tolerance and understanding among people. Today, the tantalizing question for science is, If the human consciousness or soul does indeed continue to exist well past the traditional marker that defines death, does it really ever die as an entity? Our new studies will continue to explore this and other significant ethical questions. For now, though, we can be certain that we humans no longer need to fear death.

ACKNOWLEDGMENTS

NOTHING IS EVER ACCOMPLISHED without countless people's help. I would thus like to sincerely thank my agent, Andrew Stuart; my editor, Roger Freet; and writers Josh Young and Steve Volk. I am also indebted to all those who have patiently taught me throughout my life whether in medicine or otherwise, especially my parents, and in particular my mother. Many thanks also to my wife, Lisa, for her unwavering support, and Dr. Ebby Elahi for his kind and continued help with this book as well as the research over many years.

AWARE STUDY COLLABORATORS

A SPECIAL THANK-YOU TO ALL the countless researchers who have participated in and helped with the AWARE study since its inception more than ten years ago. Particular thanks go to Professors Stephen Holgate, Robert Peveler, and Paul Little, who have supported the study at the University of Southampton at different periods, as well as Drs. Peter Fenwick and Ken Spearpoint, who have been especially involved with the development of this study. Although there are many more people who have helped enormously, I am able to provide only the names of the principal personnel and investigators at each of the participating hospitals (outside Stony Brook Medical Center in New York).

United Kingdom

University of Southampton: Professor Charles Deakin (cardiac anesthesia), Professor Paul Little (research design), Professor Robert Peveler (psychiatry), Professor Stephen Holgate (pulmonary), Ms. Katie Baker (resuscitation), Ms. Niki Fallowfield (resuscitation), Ms. Kayla Harris (resuscitation)
Addenbrook's Hospital, Cambridge: Ms. Susan Jones (resuscitation)
Northampton Hospital: Ms. Celia Warlow (resuscitation), Ms. Siobhan O'Donoghue (resuscitation)
St. Georges Hospital, London: Ms. Paula McLean (resuscitation)
St. Peters Hospital: Mr. Paul Wills (resuscitation)

Mayday Hospital, London: Mr. Russell Metcalfe Smith (resuscitation)
Royal Bournemouth Hospital: Ms. Hayley Killingback (resuscitation)
Stevenage Hospital: Ms. Salli Lovett (critical care)
Salisbury Hospital: Mr. Iain Macleod (resuscitation)
James Paget Hospital: Ms. Pam Cushing (resuscitation)
East Sussex Hospitals: Dr. Harry Walmsley (anesthetics and resuscitation)
Hammersmith Hospital, London: Dr. Ken Spearpoint (resuscitation)

United States

Indiana State University: Dr. Mark Faber (pulmonary and critical care)
University of Virginia: Professor Bruce Greyson (psychiatry), Dr. Robert O'Connor (emergency)
Emory Medical Center: Dr. Maziar Zafari (cardiology)

Austria

University of Vienna: Professor Roland Beisteiner (neurology), Dr. Fritz Sterz (emergency medicine), Dr. Michael Berger (neuroscience)

BIBLIOGRAPHY AND RESOURCES FOR
FURTHER READING

T HIS BIBLIOGRAPHY HAS BEEN provided to assist readers with a deeper interest in some of the topics covered in the book. The references relate to the main subjects under discussion and have been broadly divided into the following four sections: (1) near-death experiences/actual-death experiences; (2) brain, consciousness, and soul; (3) brain death, minimally conscious state, and vegetative state; and (4) advances in cardiac arrest and resuscitation. (Although many of these references were used in the book, I have included additional references to provide further material for deeper inquiry.)

Advances in Cardiac Arrest and Resuscitation

Abella, B. S. (2008). Hypothermia and coronary intervention after cardiac arrest: thawing a cool relationship? *Crit Care Med, 36*(6), 1967–1968.

Abella, B. S., Alvarado, J. P., Myklebust, H., Edelson, D. P., Barry, A., O'Hearn, N., . . . Becker, L. B. (2005). Quality of cardiopulmonary resuscitation during in-hospital cardiac arrest. *JAMA, 293*(3), 305–310.

Abella, B. S., Edelson, D. P., Kim, S., Retzer, E., Myklebust, H., Barry, A. M., . . . Becker, L. B. (2007). CPR quality improvement during in-hospital cardiac arrest using a real-time audiovisual feedback system. *Resuscitation, 73*(1), 54–61.

Abella, B. S., Rhee, J. W., Huang, K. N., Vanden Hoek, T. L., & Becker, L. B. (2005). Induced hypothermia is underused after resuscitation from cardiac arrest: a current practice survey. *Resuscitation, 64*(2), 181–186.

Abella, B. S., Sandbo, N., Vassilatos, P., Alvarado, J. P., O'Hearn, N., Wigder, H. N., . . . Becker, L. B. (2005). Chest compression rates during cardiopulmonary resuscitation are suboptimal: a prospective study during in-hospital cardiac arrest. *Circulation, 111*(4), 428–434.

Adrie, C., Adib-Conquy, M., Laurent, I., Monchi, M., Vinsonneau, C., Fitting, C., . . . Cavaillon, J. M. (2002). Successful cardiopulmonary resuscitation after cardiac arrest as a "sepsis-like" syndrome. *Circulation, 106*(5), 562–568.

Adrie, C., Haouache, H., Saleh, M., Memain, N., Laurent, I., Thuong, M., . . . Monchi, M. (2008). An underrecognized source of organ donors: patients with brain death after successfully resuscitated cardiac arrest. *Intensive Care Med, 34*(1), 132–137.

Alexander, M. P., Lafleche, G., Schnyer, D., Lim, C., & Verfaellie, M. (2011). Cognitive and functional outcome after out of hospital cardiac arrest. *J Int Neuropsychol Soc, 17*(2), 364–368.

Alfonzo, A., Lomas, A., Drummond, I., & McGugan, E. (2009). Survival after 5-h resuscitation attempt for hypothermic cardiac arrest using CVVH for extracorporeal rewarming. *Nephrol Dial Transplant, 24*(3), 1054–1056.

Alsoufi, B., Al-Radi, O. O., Nazer, R. I., Gruenwald, C., Foreman, C., Williams, W. G., . . . Van Arsdell, G. S. (2007). Survival outcomes after rescue extracorporeal cardiopulmonary resuscitation in pediatric patients with refractory cardiac arrest. *J Thorac Cardiovasc Surg, 134*(4), 952–959.

Andreka, P., & Frenneaux, M. P. (2006). Haemodynamics of cardiac arrest and resuscitation. *Curr Opin Crit Care, 12*(3), 198–203. Angelos, M. G., Yeh, S. T., & Aune, S. E. (2011). Post-cardiac arrest hyperoxia and mitochondrial function. *Resuscitation, 82*(Suppl 2), S48–51.

Angelos, M., Safar, P., & Reich, H. (1991). A comparison of cardiopulmonary resuscitation with cardiopulmonary bypass after prolonged cardiac arrest in dogs. Reperfusion pressures and neurologic recovery. *Resuscitation, 21*(2–3), 121–135.

Anyfantakis, Z. A., Baron, G., Aubry, P., Himbert, D., Feldman, L. J., Juliard, J. M., . . . Steg, P. G. (2009). Acute coronary angiographic findings in survivors of out-of-hospital cardiac arrest. *Am Heart J, 157*(2), 312–318.

Apps, A., Malhotra, A., Mason, M., & Lane, R. (2012). Regional systems of care after out-of-hospital cardiac arrest in the UK: premier league care saves lives. *J R Soc Med, 105*(9), 362–364.

Arrich, J., & European Resuscitation Council Hypothermia After Cardiac Arrest Registry Study Group. (2007). Clinical application of mild therapeutic hypothermia after cardiac arrest. *Crit Care Med, 35*(4), 1041–1047.

Aufderheide, T. P. (2006). The problem with and benefit of ventilations: should our approach be the same in cardiac and respiratory arrest? *Curr Opin Crit Care, 12*(3), 207–212.

Aufderheide, T. P., Alexander, C., Lick, C., Myers, B., Romig, L., Vartanian, L., . . . Lurie, K. (2008). From laboratory science to six emergency medical services systems: new understanding of the physiology of cardiopulmonary resuscitation increases survival rates after cardiac arrest. *Crit Care Med, 36*(Suppl 11), S397–404.

Aufderheide, T. P., Frascone, R. J., Wayne, M. A., Mahoney, B. D., Swor, R. A., Domeier, R. M., . . . Lurie, K. G. (2011). Standard cardiopulmonary resuscitation versus active compression-decompression cardiopulmonary resuscitation with augmentation of negative intrathoracic pressure for out-of-hospital cardiac arrest: a randomised trial. *Lancet, 377*(9762), 301–311.

Aufderheide, T. P., Kudenchuk, P. J., Hedges, J. R., Nichol, G., Kerber, R. E., Dorian, P., . . . Resuscitation Outcomes Consortium Investigators. (2008). Resuscitation Outcomes Consortium (ROC) PRIMED cardiac arrest trial methods part 1: rationale and methodology for the impedance threshold device (ITD) protocol. *Resuscitation, 78*(2), 179–185.

Aufderheide, T. P., & Lurie, K. G. (2004). Death by hyperventilation: a common and life-threatening problem during cardiopulmonary resuscitation. *Crit Care Med, 32*(9 Suppl), S345–351.

Aufderheide, T. P., Nichol, G., Rea, T. D., Brown, S. P., Leroux, B. G., Pepe, P. E., . . . Resuscitation Outcomes Consortium Investigators. (2011). A trial of an impedance threshold device in out-of-hospital cardiac arrest. *N Engl J Med, 365*(9), 798–806.

Aufderheide, T. P., Pirrallo, R. G., Provo, T. A., & Lurie, K. G. (2005). Clinical evaluation of an inspiratory impedance threshold device during standard cardiopulmonary resuscitation in patients with out-of-hospital cardiac arrest. *Crit Care Med, 33*(4), 734–740.

Aufderheide, T. P., Sigurdsson, G., Pirrallo, R. G., Yannopoulos, D., McKnite, S., von Briesen, C., . . . Lurie, K. G. (2004). Hyperventilation-induced hypotension during cardiopulmonary resuscitation. *Circulation, 109*(16), 1960–1965.

Aufderheide, T. P., Yannopoulos, D., Lick, C. J., Myers, B., Romig, L. A., Stothert, J. C., . . . Benditt, D. G. (2010). Implementing the 2005 American

Heart Association Guidelines improves outcomes after out-of-hospital cardiac arrest. *Heart Rhythm, 7*(10), 1357–1362.

Australian Resuscitation Council & New Zealand Resuscitation Council. (2011). Therapeutic hypothermia after cardiac arrest. ARC and NZRC Guideline 2010. *Emerg Med Australas, 23*(3), 297–298.

Avalli, L., Maggioni, E., Formica, F., Redaelli, G., Migliari, M., Scanziani, M., ... Fumagalli, R. (2012). Favourable survival of in-hospital compared to out-of-hospital refractory cardiac arrest patients treated with extracorporeal membrane oxygenation: an Italian tertiary care centre experience. *Resuscitation, 83*(5), 579–583.

Axelsson, C., Karlsson, T., Axelsson, A. B., & Herlitz, J. (2009). Mechanical active compression-decompression cardiopulmonary resuscitation (ACD-CPR) versus manual CPR according to pressure of end tidal carbon dioxide (P(ET)CO2) during CPR in out-of-hospital cardiac arrest (OHCA). *Resuscitation, 80*(10), 1099–1103.

Axelsson, C., Nestin, J., Svensson, L., Axelsson, A. B., & Herlitz, J. (2006). Clinical consequences of the introduction of mechanical chest compression in the EMS system for treatment of out-of-hospital cardiac arrest—a pilot study. *Resuscitation, 71*(1), 47–55.

Bachman, J. W. (1984). Cardiac arrest in the community: how to improve survival rates. *Postgrad Med, 76*(3), 85–90, 92–85.

Bangalore, S., & Hochman, J. S. (2010). A routine invasive strategy for out-of-hospital cardiac arrest survivors: are we there yet? *Circ Cardiovasc Interv, 3*(3), 197–199.

Barnes, T. A. (2010). Improving survival from in-hospital cardiac arrest. *Respir Care, 55*(8), 1100–1102.

Basu, S., Liu, X., Nozari, A., Rubertsson, S., Miclescu, A., & Wiklund, L. (2003). Evidence for time-dependent maximum increase of free radical damage and eicosanoid formation in the brain as related to duration of cardiac arrest and cardio-pulmonary resuscitation. *Free Radic Res, 37*(3), 251–256.

Baumgartner, F. J., Janusz, M. T., Jamieson, W. R., Winkler, T., Burr, L. H., & Vestrup, J. A. (1992). Cardiopulmonary bypass for resuscitation of patients with accidental hypothermia and cardiac arrest. *Can J Surg, 35*(2), 184–187.

Becker, L. B., Han, B. H., Meyer, P. M., Wright, F. A., Rhodes, K. V., Smith, D. W., & Barrett, J. (1993). Racial differences in the incidence of cardiac arrest and subsequent survival. The CPR Chicago Project. *N Engl J Med, 329*(9), 600–606.

Beckers, S. K., & Fries, M. (2010). Therapeutic mild hypothermia in cardiac arrest: a history of success? *Minerva Anesthesiol, 76*(10):778–779.

Beckstead, J. E., Tweed, W. A., Lee, J., & MacKeen, W. L. (1978). Cerebral blood flow and metabolism in man following cardiac arrest. *Stroke, 9*(6), 569–573.

Beddingfield, E., & Clark, A. P. (2012). Therapeutic hypothermia after cardiac arrest: improving adherence to national guidelines. *Clin Nurse Spec, 26*(1), 12–18.

Behringer, W., Arrich, J., Holzer, M., & Sterz, F. (2009). Out-of-hospital therapeutic hypothermia in cardiac arrest victims. *Scand J Trauma Resusc Emerg Med, 17*(52). doi: 10.1186/1757-7241-17-52.

Bellomo, R., Bailey, M., Eastwood, G. M., Nichol, A., Pilcher, D., Hart, G. K., . . . Study of Oxygen in Critical Care Group. (2011). Arterial hyperoxia and in-hospital mortality after resuscitation from cardiac arrest. *Crit Care, 15*(2), R90.

Bellville, J. W., Artusio, J. F., Jr., & Glenn, F. (1955). The electroencephalogram in cardiac arrest. *JAMA, 157*(6), 508–510.

Bennett, D. R., Nord, N. M., Roberts, T. S., & Mavor, H. (1971). Prolonged "survival" with flat EEG following cardiac arrest. *Electroencephalogr Clin Neurophysiol, 30*(1), 94.

Benson, D. W., Williams, G. R., Jr., Spencer, F. C., & Yates, A. J. (1959). The use of hypothermia after cardiac arrest. *Anesth Analg, 38*, 423–428.

Berg, R. A., Sanders, A. B., Kern, K. B., Hilwig, R. W., Heidenreich, J. W., Porter, M. E., & Ewy, G. A. (2001). Adverse hemodynamic effects of interrupting chest compressions for rescue breathing during cardiopulmonary resuscitation for ventricular fibrillation cardiac arrest. *Circulation, 104*(20), 2465–2470.

Bernard, S. A. (2005). Hypothermia improves outcome from cardiac arrest. *Crit Care Resusc, 7*(4), 325–327.

Bernard, S. A., Gray, T. W., Buist, M. D., Jones, B. M., Silvester, W., Gutteridge, G., & Smith, K. (2002). Treatment of comatose survivors of out-of-hospital cardiac arrest with induced hypothermia. *N Engl J Med, 346*(8), 557–563.

Bernard, S. A., Jones, B. M., & Horne, M. K. (1997). Clinical trial of induced hypothermia in comatose survivors of out-of-hospital cardiac arrest. *Ann Emerg Med, 30*(2), 146–153.

Bevers, M. B., Ingleton, L. P., Che, D., Cole, J. T., Li, L., Da, T., . . . Neumar, R. W. (2010). RNAi targeting micro-calpain increases neuron survival and preserves hippocampal function after global brain ischemia. *Exp Neurol, 224*(1), 170–177.

Bevers, M. B., Lawrence, E., Maronski, M., Starr, N., Amesquita, M., & Neumar, R. W. (2009). Knockdown of m-calpain increases survival of primary hippocampal neurons following NMDA excitotoxicity. *J Neurochem, 108*(5), 1237–1250.

Bevers, M. B., & Neumar, R. W. (2008). Mechanistic role of calpains in postischemic neurodegeneration. *J Cereb Blood Flow Metab, 28*(4), 655–673.

Bigham, B. L., Dainty, K. N., Scales, D. C., Morrison, L. J., & Brooks, S. C. (2010). Predictors of adopting therapeutic hypothermia for post-cardiac arrest patients among Canadian emergency and critical care physicians. *Resuscitation, 81*(1), 20–24.

Binks, A. C., Murphy, R. E., Prout, R. E., Bhayani, S., Griffiths, C. A., Mitchell, T., ... Nolan, J. P. (2010). Therapeutic hypothermia after cardiac arrest-implementation in UK intensive care units. *Anaesthesia, 65*(3), 260–265.

Binks, A., & Nolan, J. P. (2010). Post-cardiac arrest syndrome. *Minerva Anesthesiol, 76*(5), 362–368.

Binnie, C. D., Lloyd, D. S., Margerison, J. H., Maynard, D., Prior, P. F., & Scott, D. F. (1970). EEG prediction of outcome after resuscitation from cardiac or respiratory arrest. *Electroencephalogr Clin Neurophysiol, 29*(1), 105.

Binnie, C. D., Prior, P. F., Lloyd, D. S., Scott, D. F., & Margerison, J. H. (1970). Electroencephalographic prediction of fatal anoxic brain damage after resuscitation from cardiac arrest. *BMJ, 4*(5730), 265–268.

Bleck, T. P. (2006). Prognostication and management of patients who are comatose after cardiac arrest. *Neurology, 67*(4), 556–557.

Blondin, N. A., & Greer, D. M. (2011). Neurologic prognosis in cardiac arrest patients treated with therapeutic hypothermia. *Neurologist, 17*(5), 241–248.

Bloom, H. L., Shukrullah, I., Cuellar, J. R., Lloyd, M. S., Dudley, S. C., Jr., & Zafari, A. M. (2007). Long-term survival after successful inhospital cardiac arrest resuscitation. *Am Heart J, 153*(5), 831–836.

Blyth, L., Atkinson, P., Gadd, K., & Lang, E. (2012). Bedside focused echocardiography as predictor of survival in cardiac arrest patients: a systematic review. *Acad Emerg Med, 19*(10): 1119–1126.

Bobrow, B. J., Vadeboncoeur, T. F., Clark, L., & Chikani, V. (2008). Establishing Arizona's statewide cardiac arrest reporting and educational network. *Prehosp Emerg Care, 12*(3), 381–387.

Bolys, R., Ingemansson, R., Sjoberg, T., & Steen, S. (1999). Vascular function in the cadaver up to six hours after cardiac arrest. *J Heart Lung Transplant, 18*(6), 582–586.

Borger van der Burg, A. E., Bax, J. J., Boersma, E., Bootsma, M., van Erven, L., van der Wall, E. E., & Schalij, M. J. (2003). Impact of percutaneous coronary intervention or coronary artery bypass grafting on outcome after nonfatal cardiac arrest outside the hospital. *Am J Cardiol, 91*(7), 785–789.

Bottiger, B. W., Schneider, A., & Popp, E. (2007). Number needed to treat = six: therapeutic hypothermia following cardiac arrest—an effective and cheap approach to save lives. *Crit Care, 11*(4), 162.

Bouch, D. C., Thompson, J. P., & Damian, M. S. (2008). Post-cardiac arrest management: more than global cooling? *Br J Anaesth, 100*(5), 591–594.

Boutilier, R. G. (2001). Mechanisms of cell survival in hypoxia and hypothermia. *J Exp Biol, 204*(Pt 18), 3171–3181.

Brady, W. J., Gurka, K. K., Mehring, B., Peberdy, M. A., O'Connor, R. E., & American Heart Association's Get with the Guidelines Investigators. (2011). In-hospital cardiac arrest: impact of monitoring and witnessed event on patient survival and neurologic status at hospital discharge. *Resuscitation, 82*(7), 845–852.

Brodersen, P. (1974). Cerebral blood flow and metabolism in coma following cardiac arrest. *Rev Electroencephalogr Neurophysiol Clin, 4*(2), 329–333.

Bro-Jeppesen, J., Kjaergaard, J., Horsted, T. I., Wanscher, M. C., Nielsen, S. L., Rasmussen, L. S., & Hassager, C. (2009). The impact of therapeutic hypothermia on neurological function and quality of life after cardiac arrest. *Resuscitation, 80*(2), 171–176.

Brooks, S. C., Bigham, B. L., & Morrison, L. J. (2011). Mechanical versus manual chest compressions for cardiac arrest. *Cochrane Database Syst Rev* (1), CD007260.

Brooks, S. C., & Morrison, L. J. (2008). Implementation of therapeutic hypothermia guidelines for post-cardiac arrest syndrome at a glacial pace: seeking guidance from the knowledge translation literature. *Resuscitation, 77*(3), 286–292.

Brooks, S. C., Schmicker, R. H., Rea, T. D., Aufderheide, T. P., Davis, D. P., Morrison, L. J., . . . Resuscitation Outcomes Consortium Investigators. (2010). Out-of-hospital cardiac arrest frequency and survival: evidence for temporal variability. *Resuscitation, 81*(2), 175–181.

Brown, C. G., Martin, D. R., Pepe, P. E., Stueven, H., Cummins, R. O., Gonzalez, E., & Jastremski, M. (1992). A comparison of standard-dose and high-dose epinephrine in cardiac arrest outside the hospital. The Multicenter High-Dose Epinephrine Study Group. *N Engl J Med, 327*(15), 1051–1055.

Brown, J. M., & Bourdeaux, C. P. (2011). Predicting neurological outcome in post cardiac arrest patients treated with hypothermia. *Resuscitation, 82*(6), 653–654.

Bruce, C. M., Reed, M. J., & Macdougall, M. (2012). Are the public ready for organ donation after out of hospital cardiac arrest? *Emerg Med J.* doi: 10.1136/emermed-2012-201135.

Buscher, H., & Nair, P. (2011). Cardiac arrest: time matters—does time of day too? *Resuscitation, 82*(6), 649–650.

Buunk, G., van der Hoeven, J. G., Frolich, M., & Meinders, A. E. (1996). Cerebral vasoconstriction in comatose patients resuscitated from a cardiac arrest? *Intensive Care Med, 22*(11), 1191–1196.

Buunk, G., van der Hoeven, J. G., & Meinders, A. E. (1997). Cerebrovascular reactivity in comatose patients resuscitated from a cardiac arrest. *Stroke, 28*(8), 1569–1573.

Buunk, G., van der Hoeven, J. G., & Meinders, A. E. (1998). A comparison of near-infrared spectroscopy and jugular bulb oximetry in comatose patients resuscitated from a cardiac arrest. *Anaesthesia, 53*(1), 13–19.

Buunk, G., van der Hoeven, J. G., & Meinders, A. E. (1999). Prognostic significance of the difference between mixed venous and jugular bulb oxygen saturation in comatose patients resuscitated from a cardiac arrest. *Resuscitation, 41*(3), 257–262.

Buunk, G., van der Hoeven, J. G., & Meinders, A. E. (2000). Cerebral blood flow after cardiac arrest. *Neth J Med, 57*(3), 106–112.

Cabanas, J. G., Brice, J. H., De Maio, V. J., Myers, B., & Hinchey, P. R. (2011). Field-induced therapeutic hypothermia for neuroprotection after out-of hospital cardiac arrest: a systematic review of the literature. *J Emerg Med, 40*(4), 400–409.

Cairns, C. B., & Niemann, J. T. (1998). Hemodynamic effects of repeated doses of epinephrine after prolonged cardiac arrest and CPR: preliminary observations in an animal model. *Resuscitation, 36*(3), 181–185.

Calhoun, C. L., & Ettinger, M. G. (1966). Unusual EEG in coma after cardiac arrest. *Electroencephalogr Clin Neurophysiol, 21*(4), 385–388.

Callaway, C. W. (2012). Induced hypothermia after cardiac arrest improves cardiogenic shock. *Crit Care Med, 40*(6), 1963–1964.

Canadian Agency for Drugs and Technologies in Health. (2010). Vasopressin as first-line therapy for cardiac arrest: a review of the guidelines and clinical-effectiveness. *CADTH Technol Overv, 1*(2), e0112.

Canadian Association of Emergency Physicians & CAEP Critical Care Committee. (2006). Guidelines for the use of hypothermia after cardiac arrest. *CJEM, 8*(2), 106–108.

Cantoni, A., & Bocchialini, C. (1962). [The circulatory action of artificial respiration during cardiac arrest: experimental contribution]. *Minerva Anesthesiol, 28*, 61–68.

Cantrell, C. L., Jr., Hubble, M. W., & Richards, M. E. (2012). Impact of delayed and infrequent administration of vasopressors on return of spontaneous circulation during out-of-hospital cardiac arrest. *Prehosp Emerg Care.* [epub ahead of print]

Carbonell, J., Carrascosa, R., Dierssen, G., Obrador, S., Oliveros, J. C., & Sevillano, M. (1963). Some electrophysiological observations in a case of deep coma secondary to cardiac arrest. *Electroencephalogr Clin Neurophysiol, 15*(3), 520–525.

Carbutti, G., Romand, J. A., Carballo, J. S., Bendjelid, S. M., Suter, P. M., & Bendjelid, K. (2003). Transcranial Doppler: an early predictor of ischemic stroke after cardiac arrest? *Anesth Analg, 97*(5), 1262–1265.

Cardarelli, M. G., Young, A. J., & Griffith, B. (2009). Use of extracorporeal membrane oxygenation for adults in cardiac arrest (E-CPR): a meta-analysis of observational studies. *ASAIO J, 55*(6), 581–586.

Cariou, A., Claessens, Y. E., Pene, F., Marx, J. S., Spaulding, C., Hababou, C., . . . Hermine, O. (2008). Early high-dose erythropoietin therapy and hypothermia after out-of-hospital cardiac arrest: a matched control study. *Resuscitation, 76*(3), 397–404.

Carr, B. G., Goyal, M., Band, R. A., Gaieski, D. F., Abella, B. S., Merchant, R. M., . . . Neumar, R. W. (2009). A national analysis of the relationship between hospital factors and post-cardiac arrest mortality. *Intensive Care Med, 35*(3), 505–511.

Carr, B. G., Kahn, J. M., Merchant, R. M., Kramer, A. A., & Neumar, R. W. (2009). Inter-hospital variability in post-cardiac arrest mortality. *Resuscitation, 80*(1), 30–34.

Casner, M., Andersen, D., & Isaacs, S. M. (2005). The impact of a new CPR assist device on rate of return of spontaneous circulation in out-of-hospital cardiac arrest. *Prehosp Emerg Care, 9*(1), 61–67.

Castrejon, S., Cortes, M., Salto, M. L., Benittez, L. C., Rubio, R., Juarez, M., . . . Fernandez Aviles, F. (2009). Improved prognosis after using mild hypothermia to treat cardiorespiratory arrest due to a cardiac cause: comparison with a control group. *Rev Esp Cardiol, 62*(7), 733–741.

Celik, T., Iyisoy, A., Yuksel, U. C., Celik, M., & Jata, B. (2010). Chill therapy in the patients with resuscitated cardiac arrest: a new weapon in the battle against anoxic brain injury. *Int J Cardiol, 138*(3), 300–302.

Cha, W. C., Lee, S. C., Shin, S. D., Song, K. J., Sung, A. J., & Hwang, S. S. (2012). Regionalisation of out-of-hospital cardiac arrest care for patients without prehospital return of spontaneous circulation. *Resuscitation, 83*(11), 1338–1342.

Chalkias, A., & Xanthos, T. (2012). Post-cardiac arrest brain injury: pathophysiology and treatment. *J Neurol Sci, 315*(1–2), 1–8.

Chalkias, A., & Xanthos, T. (2012). Redox-mediated programed death of myocardial cells after cardiac arrest and cardiopulmonary resuscitation. *Redox Rep, 17*(2), 80–83.

Chan, P. S., Krumholz, H. M., Nichol, G., Nallamothu, B. K., & American Heart Association National Registry of Cardiopulmonary Resuscitation Investigators. (2008). Delayed time to defibrillation after in-hospital cardiac arrest. *N Engl J Med, 358*(1), 9–17.

Chan, P. S., & Nallamothu, B. K. (2012). Improving outcomes following in-hospital cardiac arrest: life after death. *JAMA, 307*(18), 1917–1918.

Chan, P. S., Nichol, G., Krumholz, H. M., Spertus, J. A., Jones, P. G., Peterson, E. D., . . . American Heart Association National Registry of Cardiopulmonary Resuscitation Investigators. (2009). Racial differences in survival after in-hospital cardiac arrest. *JAMA, 302*(11), 1195–1201.

Chan, P. S., Nichol, G., Krumholz, H. M., Spertus, J. A., Nallamothu, B. K., & American Heart Association National Registry of Cardiopulmonary Resuscitation Investigators. (2009). Hospital variation in time to defibrillation after in-hospital cardiac arrest. *Arch Intern Med, 169*(14), 1265–1273.

Chen, Y. S., Chao, A., Yu, H. Y., Ko, W. J., Wu, I. H., Chen, R. J., . . . Wang, S. S. (2003). Analysis and results of prolonged resuscitation in cardiac arrest patients rescued by extracorporeal membrane oxygenation. *J Am Coll Cardiol, 41*(2), 197–203.

Chen, Y. S., Lin, J. W., Yu, H. Y., Ko, W. J., Jerng, J. S., Chang, W. T., . . . Lin, F. Y. (2008). Cardiopulmonary resuscitation with assisted extracorporeal life-support versus conventional cardiopulmonary resuscitation in adults with in-hospital cardiac arrest: an observational study and propensity analysis. *Lancet, 372*(9638), 554–561.

Cleveland, J. C. (1971). Complete recovery after cardiac arrest for three hours. *N Engl J Med, 284*(6), 334–335.

Cloostermans, M. C., van Meulen, F. B., Eertman, C. J., Hom, H. W., & van

Putten, M. J. (2012). Continuous electroencephalography monitoring for early prediction of neurological outcome in postanoxic patients after cardiac arrest: a prospective cohort study. *Crit Care Med, 40*(10), 2867–2875.

Clute, H. L., & Levy, W. J. (1990). Electroencephalographic changes during brief cardiac arrest in humans. *Anesthesiology, 73*(5), 821–825.

Cobb, L. A. (1993). Variability in resuscitation rates for out-of-hospital cardiac arrest. *Arch Intern Med, 153*(10), 1165–1166.

Cohan, S. L., Mun, S. K., Petite, J., Correia, J., Tavelra Da Silva, A. T., & Waldhorn, R. E. (1989). Cerebral blood flow in humans following resuscitation from cardiac arrest. *Stroke, 20*(6), 761–765.

Conseil francais de reanimation cardiopulmonaire, Societe francaise d'anesthesie et de reanimation, Societe francaise de cardiologie, Societe francaise de chirurgie thoracique et cardiovasculaire, Societe francaise de medecine d'urgence, Societe francaise de pediatrie, . . . Societe de reanimation de langue francaise. (2009). Guidelines for indications for the use of extracorporeal life support in refractory cardiac arrest. French Ministry of Health. *Ann Fr Anesth Reanim, 28*(2), 182–190.

Cooper, J. A., Cooper, J. D., & Cooper, J. M. (2006). Cardiopulmonary resuscitation: history, current practice, and future direction. *Circulation, 114*(25), 2839–2849.

Cour, M., Loufouat, J., Paillard, M., Augeul, L., Goudable, J., Ovize, M., & Argaud, L. (2011). Inhibition of mitochondrial permeability transition to prevent the post-cardiac arrest syndrome: a pre-clinical study. *Eur Heart J, 32*(2), 226–235.

Cowie, M. R., Fahrenbruch, C. E., Cobb, L. A., & Hallstrom, A. P. (1993). Out-of-hospital cardiac arrest: racial differences in outcome in Seattle. *Am J Public Health, 83*(7), 955–959.

Criley, J. M., Niemann, J. T., & Rosborough, J. P. (1984). Cardiopulmonary resuscitation research 1960–1984: discoveries and advances. *Ann Emerg Med, 13*(9 Pt 2), 756–758.

Cronberg, T., Lilja, G., Rundgren, M., Friberg, H., & Widner, H. (2009). Long-term neurological outcome after cardiac arrest and therapeutic hypothermia. *Resuscitation, 80*(10), 1119–1123.

Cronberg, T., Rundgren, M., Westhall, E., Englund, E., Siemund, R., Rosen, I., . . . Friberg, H. (2011). Neuron-specific enolase correlates with other prognostic markers after cardiac arrest. *Neurology, 77*(7), 623–630.

Cronberg, T., Wise, M. P., & Nielsen, N. (2012). Mild-induced hypothermia and neuroprognostication following cardiac arrest. *Crit Care Med, 40*(8), 2537–2538; author reply 2538–2539.

Crow, H. J., & Winter, A. (1969). Serial electrophysiological studies (EEG, EMG, ERG, evoked responses) in a case of 3 months' survival with flat EEG following cardiac arrest. *Electroencephalogr Clin Neurophysiol, 27*(3), 332–333.

Cummings, B., Noviski, N., Moreland, M. P., & Paris, J. J. (2009). Circulatory arrest in a brain-dead organ donor: is the use of cardiac compression permissible? *J Intensive Care Med, 24*(6), 389–392.

Cummins, R. O., Ornato, J. P., Thies, W. H., & Pepe, P. E. (1991). Improving survival from sudden cardiac arrest: the "chain of survival" concept: a statement for health professionals from the Advanced Cardiac Life Support Subcommittee and the Emergency Cardiac Care Committee, American Heart Association. *Circulation, 83*(5), 1832–1847.

Davis, D. P. (2010). The ART of resuscitation: a new program for cardiopulmonary arrest calls. *JEMS, 35*(9), 48–49.

Deakin, C. D., Morrison, L. J., Morley, P. T., Callaway, C. W., Kerber, R. E., Kronick, S. L., . . . Advanced Life Support Chapter Collaborators. (2010). Part 8: advanced life support: 2010 International Consensus on Cardiopulmonary Resuscitation and Emergency Cardiovascular Care Science with Treatment Recommendations. *Resuscitation, 81*(Suppl 1), e93–174.

De Georgia, M. (2004). Hypothermia during cardiac arrest: moving from defense to offense. *Crit Care Med, 32*(10), 2164–2165.

De Georgia, M., & Raad, B. (2012). Prognosis of coma after cardiac arrest in the era of hypothermia. *Continuum (Minneap Minn), 18*(3), 515–531.

Delmo Walter, E. M., Alexi-Meskishvili, V., Huebler, M., Redlin, M., Boettcher, W., Weng, Y., . . . Hetzer, R. (2011). Rescue extracorporeal membrane oxygenation in children with refractory cardiac arrest. *Interact Cardiovasc Thorac Surg, 12*(6), 929–934.

De Maio, V. J. (2005). The quest to improve cardiac arrest survival: overcoming the hemodynamic effects of ventilation. *Crit Care Med, 33*(4), 898–899.

Dezfulian, C., Shiva, S., Alekseyenko, A., Pendyal, A., Beiser, D. G., Munasinghe, J. P., . . . Gladwin, M. T. (2009). Nitrite therapy after cardiac arrest reduces reactive oxygen species generation, improves cardiac and neurological function, and enhances survival via reversible inhibition of mitochondrial complex I. *Circulation, 120*(10), 897–905.

Dhanani, S., Hornby, L., Ward, R., & Shemie, S. (2012). Variability in the determination of death after cardiac arrest: a review of guidelines and statements. *J Intensive Care Med, 27*(4), 238–252.

Dhanani, S., Ward, R., Hornby, L., Barrowman, N. J., Hornby, K., Shemie,

S. D., . . . Tissue Donation. (2012). Survey of determination of death after cardiac arrest by intensive care physicians. *Crit Care Med, 40*(5), 1449–1455.

Dickinson, E. T., Verdile, V. P., Schneider, R. M., & Salluzzo, R. F. (1998). Effectiveness of mechanical versus manual chest compressions in out-of-hospital cardiac arrest resuscitation: a pilot study. *Am J Emerg Med, 16*(3), 289–292.

Don, C. W., Longstreth, W. T., Jr., Maynard, C., Olsufka, M., Nichol, G., Ray, T., . . . Kim, F. (2009). Active surface cooling protocol to induce mild therapeutic hypothermia after out-of-hospital cardiac arrest: a retrospective before-and-after comparison in a single hospital. *Crit Care Med, 37*(12), 3062–3069.

Donadello, K., Favory, R., Salgado-Ribeiro, D., Vincent, J. L., Gottin, L., Scolletta, S., . . . Taccone, F. S. (2011). Sublingual and muscular microcirculatory alterations after cardiac arrest: a pilot study. *Resuscitation, 82*(6), 690–695.

Donatelli, L. A., Geocadin, R. G., & Williams, M. A. (2006). Ethical issues in critical care and cardiac arrest: clinical research, brain death, and organ donation. *Semin Neurol, 26*(4), 452–459.

Donnino, M. W., Miller, J. C., Bivens, M., Cocchi, M. N., Salciccioli, J. D., Farris, S., . . . Howell, M. (2012). A pilot study examining the severity and outcome of the post-cardiac arrest syndrome: a comparative analysis of two geographically distinct hospitals. *Circulation, 126*(12), 1478–1483.

Donnino, M. W., Miller, J., Goyal, N., Loomba, M., Sankey, S. S., Dolcourt, B., . . . Wira, C. (2007). Effective lactate clearance is associated with improved outcome in post-cardiac arrest patients. *Resuscitation, 75*(2), 229–234.

Donnino, M. W., Rittenberger, J. C., Gaieski, D., Cocchi, M. N., Giberson, B., Peberdy, M. A., . . . Callaway, C. (2011). The development and implementation of cardiac arrest centers. *Resuscitation, 82*(8), 974–978.

Druss, R. G., & Kornfeld, D. S. (1967). The survivors of cardiac arrest: a psychiatric study. *JAMA, 201*(5), 291–296.

Drysdale, E. E., Grubb, N. R., Fox, K. A., & O'Carroll, R. E. (2000). Chronicity of memory impairment in long-term out-of-hospital cardiac arrest survivors. *Resuscitation, 47*(1), 27–32.

Duchateau, F. X., Gueye, P., Curac, S., Tubach, F., Broche, C., Plaisance, P., . . . Ricard-Hibon, A. (2010). Effect of the AutoPulse automated band chest compression device on hemodynamics in out-of-hospital cardiac arrest resuscitation. *Intensive Care Med, 36*(7), 1256–1260.

Duff, J. P., Joffe, A. R., Sevcik, W., & deCaen, A. (2011). Autoresuscitation after pediatric cardiac arrest: is hyperventilation a cause? *Pediatr Emerg Care, 27*(3), 208–209.

Dumas, F., Cariou, A., Manzo-Silberman, S., Grimaldi, D.,Vivien, B., Rosencher, J., . . . Spaulding, C. (2010). Immediate percutaneous coronary intervention is associated with better survival after out-of-hospital cardiac arrest: insights from the PROCAT (Parisian Region Out of Hospital Cardiac Arrest) registry. *Circ Cardiovasc Interv, 3*(3), 200–207.

Dumas, F., Grimaldi, D., Zuber, B., Fichet, J., Charpentier, J., Pene, F., . . . Cariou, A. (2011). Is hypothermia after cardiac arrest effective in both shockable and nonshockable patients? Insights from a large registry. *Circulation, 123*(8), 877–886.

Dumas, F., Manzo-Silberman, S., Fichet, J., Mami, Z., Zuber, B., Vivien, B., . . . Cariou,A. (2012). Can early cardiac troponin I measurement help to predict recent coronary occlusion in out-of-hospital cardiac arrest survivors? *Crit Care Med, 40*(6), 1777–1784.

Dumas, F., & Rea,T. D. (2012). Long-term prognosis following resuscitation from out-of-hospital cardiac arrest: role of aetiology and presenting arrest rhythm. *Resuscitation, 83*(8), 1001–1005.

Dumas, F.,White, L., Stubbs, B. A., Cariou, A., & Rea, T. D. (2012). Long-term prognosis following resuscitation from out of hospital cardiac arrest: role of percutaneous coronary intervention and therapeutic hypothermia. *J Am Coll Cardiol, 60*(1), 21–27.

Dunne, R. B., Compton, S., Zalenski, R. J., Swor, R., Welch, R., & Bock, B. F. (2007). Outcomes from out-of-hospital cardiac arrest in Detroit. *Resuscitation, 72*(1), 59–65.

Eapen, Z. J., Peterson, E. D., Fonarow, G. C., Sanders, G. D.,Yancy, C. W., Sears, S. F., Jr., . . . Al-Khatib, S. M. (2011). Quality of care for sudden cardiac arrest: proposed steps to improve the translation of evidence into practice. *Am Heart J, 162*(2), 222–231.

Earnest, M. P.,Yarnell, P. R., Merrill, S. L., & Knapp, G. L. (1980). Long-term survival and neurologic status after resuscitation from out-of-hospital cardiac arrest. *Neurology, 30*(12), 1298–1302.

Eckstein, M., Hatch, L., Malleck, J., McClung, C., & Henderson, S. O. (2011). End-tidal CO_2 as a predictor of survival in out-of-hospital cardiac arrest. *Prehosp Disaster Med, 26*(3), 148–150.

Eckstein, M., Stratton, S. J., & Chan, L. S. (2005). Cardiac Arrest Resuscitation Evaluation in Los Angeles: CARE-LA. *Ann Emerg Med, 45*(5), 504–509.

Edelson, D. P., Abella, B. S., Kramer-Johansen, J., Wik, L., Myklebust, H., Barry, A. M., . . . Becker, L. B. (2006). Effects of compression depth and pre-shock pauses predict defibrillation failure during cardiac arrest. *Resuscitation, 71*(2), 137–145.

Edelson, D. P., Litzinger, B., Arora, V., Walsh, D., Kim, S., Lauderdale, D. S., . . . Abella, B. S. (2008). Improving in-hospital cardiac arrest process and outcomes with performance debriefing. *Arch Intern Med, 168*(10), 1063–1069.

Edgren, E., Hedstrand, U., Kelsey, S., Sutton-Tyrrell, K., & Safar, P. (1994). Assessment of neurological prognosis in comatose survivors of cardiac arrest. BRCT I Study Group. *Lancet, 343*(8905), 1055–1059.

Edgren, E., Hedstrand, U., Nordin, M., Rydin, E., & Ronquist, G. (1987). Prediction of outcome after cardiac arrest. *Crit Care Med, 15*(9), 820–825.

Egan, A. G., & Reid, J. J. (1986). Cardiac arrest and heart attack: an evaluation of lay knowledge. *N Z Med J, 99*(799), 237–240.

Eisenberg, M. S. (2007). Improving survival from out-of-hospital cardiac arrest: back to the basics. *Ann Emerg Med, 49*(3), 314–316.

Eisenberg, M. S., Copass, M. K., Hallstrom, A., Cobb, L. A., & Bergner, L. (1980). Management of out-of-hospital cardiac arrest: failure of basic emergency medical technician services. *JAMA, 243*(10), 1049–1051.

Eisenberg, M. S., Hallstrom, A., & Bergner, L. (1982). Long-term survival after out-of-hospital cardiac arrest. *N Engl J Med, 306*(22), 1340–1343.

Eisenberg, M. S., Horwood, B. T., Cummins, R. O., Reynolds-Haertle, R., & Hearne, T. R. (1990). Cardiac arrest and resuscitation: a tale of 29 cities. *Ann Emerg Med, 19*(2), 179–186.

Eisenberg, M. S., & Psaty, B. M. (2009). Defining and improving survival rates from cardiac arrest in US communities. *JAMA, 301*(8), 860–862.

Eisenberg, M., & White, R. D. (2009). The unacceptable disparity in cardiac arrest survival among American communities. *Ann Emerg Med, 54*(2), 258–260.

Eisenburger, P., Sterz, F., Holzer, M., Zeiner, A., Scheinecker, W., Havel, C., & Losert, H. (2001). Therapeutic hypothermia after cardiac arrest. *Curr Opin Crit Care, 7*(3), 184–188.

Ekstrom, L., Herlitz, J., Wennerblom, B., Axelsson, A., Bang, A., & Holmberg, S. (1994). Survival after cardiac arrest outside hospital over a 12-year period in Gothenburg. *Resuscitation, 27*(3), 181–187.

Elliott, V. J., Rodgers, D. L., & Brett, S. J. (2011). Systematic review of quality of life and other patient-centred outcomes after cardiac arrest survival. *Resuscitation, 82*(3), 247–256.

Emergency treatment for cardiac arrest falls short of guidelines. (2005). *Heart Advis, 8*(3), 2.

Engdahl, J. (2002). Outcome after cardiac arrest outside hospital. *BMJ, 325*(7363), 503–504.

Engdahl, J., Bang, A., Lindqvist, J., & Herlitz, J. (2001). Factors affecting short- and long-term prognosis among 1069 patients with out-of-hospital cardiac arrest and pulseless electrical activity. *Resuscitation, 51*(1), 17–25.

Fabbri, L. P., Nucera, M., Becucci, A., Grippo, A., Venneri, F., Merciai, V., & Boncinelli, S. (2001). An exceptional case of complete neurologic recovery after more than 5-h cardiac arrest. *Resuscitation, 48*(2), 175–180.

Fairbanks, R. J., Shah, M. N., Lerner, E. B., Ilangovan, K., Pennington, E. C., & Schneider, S. M. (2007). Epidemiology and outcomes of out-of-hospital cardiac arrest in Rochester, New York. *Resuscitation, 72*(3), 415–424.

Fauvage, B., & Combes, P. (1993). Isoelectric electroencephalogram and loss of evoked potentials in a patient who survived cardiac arrest. *Crit Care Med, 21*(3), 472–475.

Feldman, E., Rubin, B., & Surks, S. N. (1960). Beneficial effects of hypothermia after cardiac arrest. *JAMA, 173*(5), 499–501.

Feng, L., Yuan, Y., Dong, J. T., Han, Y., Deng, Z. H., Liu, W. Q., & Li, B. F. (2012). Extracorporeal membrane oxygenation support in resuscitations for acute myocardial infarction with cardiac arrest. *Chin Med Sci J, 27*(1), 60–61.

Ferguson, L. P., Durward, A., & Tibby, S. M. (2012). Relationship between arterial partial oxygen pressure after resuscitation from cardiac arrest and mortality in children. *Circulation, 126*(3), 335–342.

Ferreira, I., Schutte, M., Oosterloo, E., Dekker, W., Mooi, B. W., Dambrink, J. H., & van 't Hof, A. W. (2009). Therapeutic mild hypothermia improves outcome after out-of-hospital cardiac arrest. *Neth Heart J, 17*(10), 378–384.

Field, J. M., Hazinski, M. F., Sayre, M. R., Chameides, L., Schexnayder, S. M., Hemphill, R., . . . Vanden Hoek, T. L. (2010). Part 1: executive summary: 2010 American Heart Association Guidelines for Cardiopulmonary Resuscitation and Emergency Cardiovascular Care. *Circulation, 122*(18 Suppl 3), S640–656.

Fink, E. L., Callaway, C. W., Tisherman, S. A., & Kochanek, P. M. (2007). Winning the cold war: inroads into implementation of mild hypothermia after cardiac arrest in adults from the European Resuscitation Council Hypothermia After Cardiac Arrest Registry Study Group. *Crit Care Med, 35*(4), 1199–1202.

Fiser, R. T., & Morris, M. C. (2008). Extracorporeal cardiopulmonary resuscitation in refractory pediatric cardiac arrest. *Pediatr Clin North Am, 55*(4), 929–941.

Frederick, J. R., Chen, Z., Bevers, M. B., Ingleton, L. P., Ma, M., & Neumar, R. W. (2008). Neuroprotection with delayed calpain inhibition after transient forebrain ischemia. *Crit Care Med, 36*(11 Suppl), S481–485.

Fries, M., Stoppe, C., Brucken, D., Rossaint, R., & Kuhlen, R. (2009). Influence of mild therapeutic hypothermia on the inflammatory response after successful resuscitation from cardiac arrest. *J Crit Care, 24*(3), 453–457. doi: 10.1016/j.jcrc.2008.10.012.

Frontera, J. A. (2012). Moving beyond moderate therapeutic hypothermia for cardiac arrest. *Crit Care Med, 40*(4), 1383–1384.

Fugate, J. E., Wijdicks, E. F., White, R. D., & Rabinstein, A. A. (2011). Does therapeutic hypothermia affect time to awakening in cardiac arrest survivors? *Neurology, 77*(14), 1346–1350.

Fujioka, M., Okuchi, K., Sakaki, T., Hiramatsu, K., Miyamoto, S., & Iwasaki, S. (1994). Specific changes in human brain following reperfusion after cardiac arrest. *Stroke, 25*(10), 2091–2095.

Gaieski, D. F., Band, R. A., Abella, B. S., Neumar, R. W., Fuchs, B. D., Kolansky, D. M., . . . Goyal, M. (2009). Early goal-directed hemodynamic optimization combined with therapeutic hypothermia in comatose survivors of out-of-hospital cardiac arrest. *Resuscitation, 80*(4), 418–424.

Gaieski, D. F., Boller, M., & Becker, L. B. (2012). Emergency cardiopulmonary bypass: a promising rescue strategy for refractory cardiac arrest. *Crit Care Clin, 28*(2), 211–229.

Gaieski, D. F., & Goyal, M. (2010). History and current trends in sudden cardiac arrest and resuscitation in adults. *Hosp Pract (Minneap), 38*(4), 44–53.

Galanaud, D., & Puybasset, L. (2010). Cardiac arrest: has the time of MRI come? *Crit Care, 14*(2), 135. doi: 10.1186/cc8905.

Gamper, G., Willeit, M., Sterz, F., Herkner, H., Zoufaly, A., Hornik, K., . . . Laggner, A. N. (2004). Life after death: posttraumatic stress disorder in survivors of cardiac arrest—prevalence, associated factors, and the influence of sedation and analgesia. *Crit Care Med, 32*(2), 378–383.

Gando, S., Igarashi, M., Kameue, T., & Nanzaki, S. (1997). Ionized hypocalcemia during out-of-hospital cardiac arrest and cardiopulmonary resuscitation is not due to binding by lactate. *Intensive Care Med, 23*(12), 1245–1250.

Garrett, J. S., Studnek, J. R., Blackwell, T., Vandeventer, S., Pearson, D. A., Heffner, A. C., & Reades, R. (2011). The association between intra-arrest therapeutic hypothermia and return of spontaneous circulation among individuals experiencing out of hospital cardiac arrest. *Resuscitation, 82*(1), 21–25.

Gates, S., Smith, J. L., Ong, G. J., Brace, S. J., & Perkins, G. D. (2012). Effectiveness of the LUCAS device for mechanical chest compression after cardiac arrest: systematic review of experimental, observational and animal studies. *Heart, 98*(12), 908–913.

Geocadin, R. G., Buitrago, M. M., Torbey, M. T., Chandra-Strobos, N., Williams, M. A., & Kaplan, P. W. (2006). Neurologic prognosis and withdrawal of life support after resuscitation from cardiac arrest. *Neurology, 67*(1), 105–108.

Geocadin, R. G., & Eleff, S. M. (2008). Cardiac arrest resuscitation: neurologic prognostication and brain death. *Curr Opin Crit Care, 14*(3), 261–268.

Geocadin, R. G., Koenig, M. A., Jia, X., Stevens, R. D., & Peberdy, M. A. (2008). Management of brain injury after resuscitation from cardiac arrest. *Neurol Clin, 26*(2), 487–506, ix.

Geocadin, R. G., Koenig, M. A., Stevens, R. D., & Peberdy, M. A. (2006). Intensive care for brain injury after cardiac arrest: therapeutic hypothermia and related neuroprotective strategies. *Crit Care Clin, 22*(4), 619–636; abstract viii.

Gillart, T., Loiseau, S., Azarnoush, K., Gonzalez, D., & Guelon, D. (2008). [Resuscitation after three hours of cardiac arrest with severe hypothermia following a toxic coma]. *Ann Fr Anesth Reanim, 27*(6), 510–513.

Ginsberg, F. L. (2011). Resuscitation from cardiac arrest: can we do better? *Crit Care Med, 39*(7), 1832–1833.

Goldberger, Z. D., Chan, P. S., Berg, R. A., Kronick, S. L., Cooke, C. R., Lu, M., . . . for the American Heart Association Get with the Guidelines—Resuscitation Investigators. (2012). Duration of resuscitation efforts and survival after in-hospital cardiac arrest: an observational study. *Lancet, 380*(9852): 1473–1481.

Graf, J., Muhlhoff, C., Doig, G. S., Reinartz, S., Bode, K., Dujardin, R., . . . Janssens, U. (2008). Health care costs, long-term survival, and quality of life following intensive care unit admission after cardiac arrest. *Crit Care, 12*(4), R92.

Granja, C., Cabral, G., Pinto, A. T., & Costa-Pereira, A. (2002). Quality of life 6-months after cardiac arrest. *Resuscitation, 55*(1), 37–44.

Grasner, J. T., Herlitz, J., Koster, R. W., Rosell-Ortiz, F., Stamatakis, L., & Bossaert, L. (2011). Quality management in resuscitation—towards a European cardiac arrest registry (EuReCa). *Resuscitation, 82*(8), 989–994. doi: 10.1016/j.resuscitation.2011.02.047.

Gratrix, A. P., Pittard, A. J., & Bodenham, A. R. (2007). Outcome after admission to ITU following out-of-hospital cardiac arrest: are non-survivors suitable for non-heart-beating organ donation? *Anaesthesia, 62*(5), 434–437. doi: 10.1111/j.1365–2044.2007.04981.x.

Grmec, S., & Klemen, P. (2001). Does the end-tidal carbon dioxide (EtCO2) concentration have prognostic value during out-of-hospital cardiac arrest? *Eur J Emerg Med, 8*(4), 263–269.

Grmec, S., Krizmaric, M., Mally, S., Kozelj, A., Spindler, M., & Lesnik, B. (2007). Utstein style analysis of out-of-hospital cardiac arrest—bystander CPR and end expired carbon dioxide. *Resuscitation, 72*(3), 404–414.

Grogaard, H. K., Wik, L., Eriksen, M., Brekke, M., & Sunde, K. (2007). Continuous mechanical chest compressions during cardiac arrest to facilitate restoration of coronary circulation with percutaneous coronary intervention. *J Am Coll Cardiol, 50*(11), 1093–1094. doi: 10.1016/j.jacc.2007.05.028.

Grubb, N. R., Elton, R. A., & Fox, K. A. (1995). In-hospital mortality after out-of-hospital cardiac arrest. *Lancet, 346*(8972), 417–421.

Grubb, N. R., Fox, K. A., Smith, K., Best, J., Blane, A., Ebmeier, K. P., . . . O'Carroll, R. E. (2000). Memory impairment in out-of-hospital cardiac arrest survivors is associated with global reduction in brain volume, not focal hippocampal injury. *Stroke, 31*(7), 1509–1514.

Grubb, N. R., O'Carroll, R., Cobbe, S. M., Sirel, J., & Fox, K. A. (1996). Chronic memory impairment after cardiac arrest outside hospital. *BMJ, 313*(7050), 143–146.

Grubb, N. R., Simpson, C., Sherwood, R. A., Abraha, H. D., Cobbe, S. M., O'Carroll, R. E., . . . Fox, K. A. (2007). Prediction of cognitive dysfunction after resuscitation from out-of-hospital cardiac arrest using serum neuron-specific enolase and protein S-100. *Heart, 93*(10), 1268–1273.

Guenther, U., Varelmann, D., Putensen, C., & Wrigge, H. (2009). Extended therapeutic hypothermia for several days during extracorporeal membrane-oxygenation after drowning and cardiac arrest: two cases of survival with no neurological sequelae. *Resuscitation, 80*(3), 379–381.

Gueugniaud, P. Y., Garcia-Darennes, F., Gaussorgues, P., Bancalari, G., Petit, P., & Robert, D. (1991). Prognostic significance of early intracranial and cerebral perfusion pressures in post-cardiac arrest anoxic coma. *Intensive Care Med, 17*(7), 392–398.

5316 Bibliography and Resources for Further Reading

Guglin, M. E., Wilson, A., Kostis, J. B., Parrillo, J. E., White, M. C., & Gessman, L. J. (2004). Immediate and 1-year survival of out-of-hospital cardiac arrest victims in southern New Jersey: 1995–2000. *Pacing Clin Electrophysiol, 27*(8), 1072–1076.

Hagiwara, S., Yamada, T., Furukawa, K., Ishihara, K., Nakamura, T., Ohyama, Y., . . . Oshima, K. (2011). Survival after 385 min of cardiopulmonary resuscitation with extracorporeal membrane oxygenation and rewarming with haemodialysis for hypothermic cardiac arrest. *Resuscitation, 82*(6), 790–791.

Hajjar, K., Kerr, D. M., & Lees, K. R. (2011). Thrombolysis for acute ischemic stroke. *J Vasc Surg, 54*(3), 901–907.

Hallstrom, A., Rea, T. D., Sayre, M. R., Christenson, J., Anton, A. R., Mosesso, V. N., Jr., . . . Cobb, L. A. (2006). Manual chest compression vs use of an automated chest compression device during resuscitation following out-of-hospital cardiac arrest: a randomized trial. *JAMA, 295*(22), 2620–2628.

Harden, A. (1969). EEG studies following resuscitation after cardiac arrest in 60 babies. *Electroencephalogr Clin Neurophysiol, 27*(3), 333.

Harden, A., Pampiglione, G., & Waterston, D. J. (1966). Circulatory arrest during hypothermia in cardiac surgery: an E.E.G. study in children. *BMJ, 2*(5522), 1105–1108.

Harkins, G. A., & Bramson, M. L. (1961). Mechanized external cardiac massage for cardiac arrest and for support of the failing heart: a preliminary communication. *J Surg Res, 1,* 197–200.

Haugk, M., Testori, C., Sterz, F., Uranitsch, M., Holzer, M., Behringer, W., . . . Time to Target Temperature Study Group. (2011). Relationship between time to target temperature and outcome in patients treated with therapeutic hypothermia after cardiac arrest. *Crit Care, 15*(2), R101.

Heart and Stroke Foundation of Canada. (2005). Therapeutic hypothermia after cardiac arrest: ILCOR Advisory Statement, October 2002. *CJEM, 7*(2), 129.

Heart attack vs. sudden cardiac arrest: what's the difference? (2004). *Harv Health Lett, 29*(10), 7.

Herlitz, J., Engdahl, J., Svensson, L., Angquist, K. A., Silfverstolpe, J., & Holmberg, S. (2006). Major differences in 1-month survival between hospitals in Sweden among initial survivors of out-of-hospital cardiac arrest. *Resuscitation, 70*(3), 404–409. doi: 10.1016/j.resuscitation.2006.01.014.

Herlitz, J., Engdahl, J., Svensson, L., Angquist, K. A., Young, M., & Holmberg, S. (2005). Factors associated with an increased chance of survival among patients suffering from an out-of-hospital cardiac arrest in a national perspective in Sweden. *Am Heart J, 149*(1), 61–66. doi: 10.1016/j.ahj.2004.07.014.

Hernandez, C., Shuler, K., Hannan, H., Sonyika, C., Likourezos, A., & Marshall, J. (2008). C.A.U.S.E.: Cardiac arrest ultra-sound exam—a better approach to managing patients in primary non-arrhythmogenic cardiac arrest. *Resuscitation, 76*(2), 198–206. doi: 10.1016/j.resuscitation.2007.06.033.

Herrera Llerandi, R., & Alvarado, G. (1959). [Cardiac arrest: treatment of acute circulatory arrest and report of a case resuscitated after 2 hours and 15 minutes]. *Rev Col Med Guatem, 10,* 196–202.

Hinchey, P. R., Myers, J. B., Lewis, R., De Maio, V. J., Reyer, E., Licatese, D., . . . Capital County Research Consortium. (2010). Improved out-of-hospital cardiac arrest survival after the sequential implementation of 2005 AHA guidelines for compressions, ventilations, and induced hypothermia: the Wake County experience. *Ann Emerg Med, 56*(4), 348–357. doi: 10.1016/j.annemergmed.2010.01.036.

Hoffmann, M., & Scherzer, E. (1970). EEG changes following temporary cardiac arrest and open cardiac massage. *Electroencephalogr Clin Neurophysiol, 28*(3), 326.

Holzer, M. (2010). Targeted temperature management for comatose survivors of cardiac arrest. *N Engl J Med, 363*(13), 1256–1264.

Holzer, M., & Behringer, W. (2005). Therapeutic hypothermia after cardiac arrest. *Curr Opin Anaesthesiol, 18*(2), 163–168.

Holzer, M., & Behringer, W. (2008). Therapeutic hypothermia after cardiac arrest and myocardial infarction. *Best Pract Res Clin Anaesthesiol, 22*(4), 711–728.

Holzer, M., Bernard, S. A., Hachimi-Idrissi, S., Roine, R. O., Sterz, F., Mullner, M., & Collaborative Group on Induced Hypothermia for Neuroprotection After Cardiac Arrest. (2005). Hypothermia for neuroprotection after cardiac arrest: systematic review and individual patient data meta-analysis. *Crit Care Med, 33*(2), 414–418.

Huang, R., Sochocka, E., & Hertz, L. (1997). Cell culture studies of the role of elevated extracellular glutamate and K+ in neuronal cell death during and after anoxia/ischemia. *Neurosci Biobehav Rev, 21*(2), 129–134.

Huang, S. C., Wu, E. T., Wang, C. C., Chen, Y. S., Chang, C. I., Chiu, I. S., . . . Wang, S. S. (2012). Eleven years of experience with extracorporeal cardiopulmonary resuscitation for paediatric patients with in-hospital cardiac arrest. *Resuscitation, 83*(6), 710–714.

Hughes, J. R., & Uppal, H. (1998). The EEG changes during cardiac arrest: a case report. *Clin Electroencephalogr, 29*(1), 16–18.

Hypothermia After Cardiac Arrest Study Group. (2002). Mild therapeutic hypothermia to improve the neurologic outcome after cardiac arrest. *N Engl J Med, 346*(8), 549–556. doi: 10.1056/NEJMoa012689.

Idris, A. H., Guffey, D., Aufderheide, T. P., Brown, S., Morrison, L. J., Nichols, P., . . . Resuscitation Outcomes Consortium Investigators. (2012). Relationship between chest compression rates and outcomes from cardiac arrest. *Circulation, 125*(24), 3004–3012.

Idris, A. H., Roberts, L. J., II, Caruso, L., Showstark, M., Layon, A. J., Becker, L. B., . . . Gabrielli, A. (2005). Oxidant injury occurs rapidly after cardiac arrest, cardiopulmonary resuscitation, and reperfusion. *Crit Care Med, 33*(9), 2043–2048.

Idris, A. H., Wenzel, V., Becker, L. B., Banner, M. J., & Orban, D. J. (1995). Does hypoxia or hypercarbia independently affect resuscitation from cardiac arrest? *Chest, 108*(2), 522–528.

Is sudden cardiac arrest the same as a heart attack? (2011). *Johns Hopkins Med Lett Health After 50, 23*(2), 8.

Italian Cooling Experience Study Group. (2012). Early- versus late-initiation of therapeutic hypothermia after cardiac arrest: preliminary observations from the experience of 17 Italian intensive care units. *Resuscitation, 83*(7), 823–828.

Ito, N., Nanto, S., Nagao, K., Hatanaka, T., & Kai, T. (2010). Regional cerebral oxygen saturation predicts poor neurological outcome in patients with out-of-hospital cardiac arrest. *Resuscitation, 81*(12), 1736–1737.

Jones, A. E. (2008). Hypothermia after cardiac arrest: we can do this. *Acad Emerg Med, 15*(6), 558–559.

Jones, A. E., Shapiro, N. I., Kilgannon, J. H., Trzeciak, S., & Emergency Medicine Shock Research Network Investigators. (2008). Goal-directed hemodynamic optimization in the post-cardiac arrest syndrome: a systematic review. *Resuscitation, 77*(1), 26–29.

Jones, K., Garg, M., Bali, D., Yang, R., & Compton, S. (2006). The knowledge and perceptions of medical personnel relating to outcome after cardiac arrest. *Resuscitation, 69*(2), 235–239.

Josephson, S. A. (2010). Predicting neurologic outcomes after cardiac arrest: the crystal ball becomes cloudy. *Ann Neurol, 67*(3), A5–6.

Kagawa, E., Dote, K., Kato, M., Sasaki, S., Nakano, Y., Kajikawa, M., . . . Kurisu, S. (2012). Should we emergently revascularize occluded coronaries for cardiac arrest? Rapid-response extracorporeal membrane oxygenation and intra-arrest percutaneous coronary intervention. *Circulation, 126*(13), 1605–1613.

Kagawa, E., Inoue, I., Kawagoe, T., Ishihara, M., Shimatani, Y., Kurisu, S., . . . Oda, N. (2010). Assessment of outcomes and differences between in- and out-of-hospital cardiac arrest patients treated with cardiopulmonary resuscitation using extracorporeal life support. *Resuscitation, 81*(8), 968–973.

Kallestedt, M. L., Rosenblad, A., Leppert, J., Herlitz, J., & Enlund, M. (2010). Hospital employees' theoretical knowledge on what to do in an in-hospital cardiac arrest. *Scand J Trauma Resusc Emerg Med, 18,* 43.

Kamarainen, A., Sainio, M., Olkkola, K. T., Huhtala, H., Tenhunen, J., & Hoppu, S. (2012). Quality controlled manual chest compressions and cerebral oxygenation during in-hospital cardiac arrest. *Resuscitation, 83*(1), 138–142.

Karanjia, N., & Geocadin, R. G. (2011). Post-cardiac arrest syndrome: update on brain injury management and prognostication. *Curr Treat Options Neurol, 13*(2), 191–203.

Kayser, R. G., Ornato, J. P., Peberdy, M. A., & American Heart Association National Registry of Cardiopulmonary Resuscitation. (2008). Cardiac arrest in the emergency department: a report from the National Registry of Cardiopulmonary Resuscitation. *Resuscitation, 78*(2), 151–160.

Kennedy, J., Green, R. S., Stenstrom, R., & CAEP Critical Care Committee. (2008). The use of induced hypothermia after cardiac arrest: a survey of Canadian emergency physicians. *CJEM, 10*(2), 125–130.

Kern, K. B., & Rahman, O. (2010). Emergent percutaneous coronary intervention for resuscitated victims of out-of-hospital cardiac arrest. *Catheter Cardiovasc Interv, 75*(4), 616–624.

Keuper, W., Dieker, H. J., Brouwer, M. A., & Verheugt, F. W. (2007). Reperfusion therapy in out-of-hospital cardiac arrest: current insights. *Resuscitation, 73*(2), 189–201.

Kilgannon, J. H., Jones, A. E., Parrillo, J. E., Dellinger, R. P., Milcarek, B., Hunter, K., . . . Emergency Medicine Shock Research Network Investigators. (2011). Relationship between supranormal oxygen tension and outcome after resuscitation from cardiac arrest. *Circulation, 123*(23), 2717–2722.

Kilgannon, J. H., Jones, A. E., Shapiro, N. I., Angelos, M. G., Milcarek, B., Hunter, K., . . . Emergency Medicine Shock Research Network Investigators. (2010). Association between arterial hyperoxia following resuscitation from cardiac arrest and in-hospital mortality. *JAMA, 303*(21), 2165–2171.

Kilgannon, J. H., Roberts, B. W., Reihl, L. R., Chansky, M. E., Jones, A. E., Dellinger, R. P., . . . Trzeciak, S. (2008). Early arterial hypotension is common in the post-cardiac arrest syndrome and associated with increased in-hospital mortality. *Resuscitation, 79*(3), 410–416.

Kim, Y. M., Yim, H. W., Jeong, S. H., Klem, M. L., & Callaway, C. W. (2012). Does therapeutic hypothermia benefit adult cardiac arrest patients presenting with non-shockable initial rhythms? A systematic review and meta-analysis of randomized and non-randomized studies. *Resuscitation, 83*(2), 188–196.

Krarup, N. H., Terkelsen, C. J., Johnsen, S. P., Clemmensen, P., Olivecrona, G. K., Hansen, T. M., . . . Lassen, J. F. (2011). Quality of cardiopulmonary resuscitation in out-of-hospital cardiac arrest is hampered by interruptions in chest compressions—a nationwide prospective feasibility study. *Resuscitation, 82*(3), 263–269.

Kremens, K., Seevaratnam, A., Fine, J., Wakefield, D. B., & Berman, L. (2011). Implementation of therapeutic hypothermia after cardiac arrest—a telephone survey of Connecticut hospitals. *Conn Med, 75*(4), 203–206.

Krnjevic, K. (1999). Early effects of hypoxia on brain cell function. *Croat Med J, 40*(3), 375–380.

Kulkens, S., & Hacke, W. (2007). Thrombolysis with alteplase for acute ischemic stroke: review of SITS-MOST and other Phase IV studies. *Expert Rev Neurother, 7*(7), 783–788.

Kumar, S., & Ewy, G. A. (2012). The hospital's role in improving survival of patients with out-of-hospital cardiac arrest. *Clin Cardiol, 35*(8), 462–466.

Larkin, G. L., Copes, W. S., Nathanson, B. H., & Kaye, W. (2010). Pre-resuscitation factors associated with mortality in 49,130 cases of in-hospital cardiac arrest: a report from the National Registry for Cardiopulmonary Resuscitation. *Resuscitation, 81*(3), 302–311.

Larsen, J. M., & Ravkilde, J. (2012). Acute coronary angiography in patients resuscitated from out-of-hospital cardiac arrest: a systematic review and meta-analysis. *Resuscitation, 83*(12), 1427–1433.

Latil, M., Rocheteau, P., Chatre, L., Sanulli, S., Memet, S., Ricchetti, M., . . . Chretien, F. (2012). Skeletal muscle stem cells adopt a dormant cell state post mortem and retain regenerative capacity. *Nat Commun, 3*, 903.

Laver, S. R., Padkin, A., Atalla, A., & Nolan, J. P. (2006). Therapeutic hypothermia after cardiac arrest: a survey of practice in intensive care units in the United Kingdom. *Anaesthesia, 61*(9), 873–877.

Le Guen, M., Nicolas-Robin, A., Carreira, S., Raux, M., Leprince, P., Riou, B., & Langeron, O. (2011). Extracorporeal life support following out-of-hospital refractory cardiac arrest. *Crit Care, 15*(1), R29.

Leung, T. W., & Wong, K. S. (2009). Thrombolysis with alteplase for acute ischemic stroke: safe and effective outside the 3-hour time window? *Nat Clin Pract Neurol, 5*(2), 70–71.

Lim, C., Alexander, M. P., LaFleche, G., Schnyer, D. M., & Verfaellie, M. (2004). The neurological and cognitive sequelae of cardiac arrest. *Neurology, 63*(10), 1774–1778.

Liu, J. M., Yang, Q., Pirrallo, R. G., Klein, J. P., & Aufderheide, T. P. (2008). Hospital variability of out-of-hospital cardiac arrest survival. *Prehosp Emerg Care, 12*(3), 339–346.

Lo, E. H. (2008). A new penumbra: transitioning from injury into repair after stroke. *Nat Med, 14*(5), 497–500.

Lombardi, G., Gallagher, J., & Gennis, P. (1994). Outcome of out-of-hospital cardiac arrest in New York City. The Pre-Hospital Arrest Survival Evaluation (PHASE) Study. *JAMA, 271*(9), 678–683.

Lundbye, J. B., Rai, M., Ramu, B., Hosseini-Khalili, A., Li, D., Slim, H. B., . . . Kluger, J. (2012). Therapeutic hypothermia is associated with improved neurologic outcome and survival in cardiac arrest survivors of non-shockable rhythms. *Resuscitation, 83*(2), 202–207.

Lund-Kordahl, I., Olasveengen, T. M., Lorem, T., Samdal, M., Wik, L., & Sunde, K. (2010). Improving outcome after out-of-hospital cardiac arrest by strengthening weak links of the local Chain of Survival: quality of advanced life support and post-resuscitation care. *Resuscitation, 81*(4), 422–426.

Lurie, K. G., Idris, A., & Holcomb, J. B. (2005). Level 1 cardiac arrest centers: learning from the trauma surgeons. *Acad Emerg Med, 12*(1), 79–80.

Lurie, K. G., Shultz, J. J., Callaham, M. L., Schwab, T. M., Gisch, T., Rector, T., . . . Long, L. (1994). Evaluation of active compression-decompression CPR in victims of out-of-hospital cardiac arrest. *JAMA, 271*(18), 1405–1411.

Martinell, L., Larsson, M., Bang, A., Karlsson, T., Lindqvist, J., Thoren, A. B., & Herlitz, J. (2010). Survival in out-of-hospital cardiac arrest before and after use of advanced postresuscitation care: a survey focusing on incidence, patient characteristics, survival, and estimated cerebral function after postresuscitation care. *Am J Emerg Med, 28*(5), 543–551.

Massetti, M., Tasle, M., Le Page, O., Deredec, R., Babatasi, G., Buklas, D., . . . Khayat, A. (2005). Back from irreversibility: extracorporeal life support for prolonged cardiac arrest. *Ann Thorac Surg, 79*(1), 178–183; discussion 183–174.

Mateen, F. J., Josephs, K. A., Trenerry, M. R., Felmlee-Devine, M. D., Weaver, A. L., Carone, M., & White, R. D. (2011). Long-term cognitive outcomes following out-of-hospital cardiac arrest: a population-based study. *Neurology*, 77(15), 1438–1445.

Mayer, S. A. (2002). Hypothermia for neuroprotection after cardiac arrest. *Curr Neurol Neurosci Rep*, 2(6), 525–526.

Mayer, S. A. (2011). Outcome prediction after cardiac arrest: new game, new rules. *Neurology*, 77(7), 614–615.

Merchant, R. M., Abella, B. S., Khan, M., Huang, K. N., Beiser, D. G., Neumar, R. W., . . . Vanden Hoek, T. L. (2008). Cardiac catheterization is underutilized after in-hospital cardiac arrest. *Resuscitation*, 79(3), 398–403.

Merchant, R. M., Becker, L. B., Abella, B. S., Asch, D. A., & Groeneveld, P. W. (2009). Cost-effectiveness of therapeutic hypothermia after cardiac arrest. *Circ Cardiovasc Qual Outcomes*, 2(5), 421–428.

Merchant, R. M., Becker, L. B., Yang, F., & Groeneveld, P. W. (2011). Hospital racial composition: a neglected factor in cardiac arrest survival disparities. *Am Heart J*, 161(4), 705–711.

Merchant, R. M., Soar, J., Skrifvars, M. B., Silfvast, T., Edelson, D. P., Ahmad, F., . . . Abella, B. S. (2006). Therapeutic hypothermia utilization among physicians after resuscitation from cardiac arrest. *Crit Care Med*, 34(7), 1935–1940.

Miglioranza, M. H., & Barbisan, J. N. (2010). Is it time for ultrasound in cardiac arrest? *J Trauma*, 68(6), 1515–1516.

Mitchison, H. M., Lim, M. J., & Cooper, J. D. (2004). Selectivity and types of cell death in the neuronal ceroid lipofuscinoses. *Brain Pathol*, 14(1), 86–96.

Morimura, N., Sakamoto, T., Nagao, K., Asai, Y., Yokota, H., Tahara, Y., . . . Hase, M. (2011). Extracorporeal cardiopulmonary resuscitation for out-of-hospital cardiac arrest: a review of the Japanese literature. *Resuscitation*, 82(1), 10–14. doi: 10.1016/j.resuscitation.2010.08.032.

Moss, J., & Rockoff, M. (1980). EEG monitoring during cardiac arrest and resuscitation. *JAMA*, 244(24), 2750–2751.

Moulaert, V. R., Verbunt, J. A., van Heugten, C. M., Bakx, W. G., Gorgels, A. P., Bekkers, S. C., . . . Wade, D. T. (2007). Activity and Life After Survival of a Cardiac Arrest (ALASCA) and the effectiveness of an early intervention service: design of a randomised controlled trial. *BMC Cardiovasc Disord*, 7, 26. doi: 10.1186/1471-2261-7-26.

Moulaert, V. R., Verbunt, J. A., van Heugten, C. M., & Wade, D. T. (2009). Cognitive impairments in survivors of out-of-hospital cardiac arrest: a systematic review. *Resuscitation*, 80(3), 297–305.

Moulaert, V. R., Wachelder, E. M., Verbunt, J. A., Wade, D. T., & van Heugten,

C. M. (2010). Determinants of quality of life in survivors of cardiac arrest. *J Rehabil Med, 42*(6), 553–558.

Mullner, M., Sterz, F., Binder, M., Hirschl, M. M., Janata, K., & Laggner, A. N. (1995). Near infrared spectroscopy during and after cardiac arrest: preliminary results. *Clin Intensive Care, 6*(3), 107–111.

Nadkarni, V. (2006). Measuring ventilation through defibrillator pads: first steps toward avoidance of "death by hyperventilation" during cardiopulmonary resuscitation? *Crit Care Med, 34*(9), 2502–2503.

Nadkarni, V. M., Larkin, G. L., Peberdy, M. A., Carey, S. M., Kaye, W., Mancini, M. E., . . . National Registry of Cardiopulmonary Resuscitation Investigators. (2006). First documented rhythm and clinical outcome from in-hospital cardiac arrest among children and adults. *JAMA, 295*(1), 50–57.

Nagao, K., Hayashi, N., Kanmatsuse, K., Arima, K., Ohtsuki, J., Kikushima, K., & Watanabe, I. (2000). Cardiopulmonary cerebral resuscitation using emergency cardiopulmonary bypass, coronary reperfusion therapy and mild hypothermia in patients with cardiac arrest outside the hospital. *J Am Coll Cardiol, 36*(3), 776–783.

Nagao, K., Kikushima, K., Watanabe, K., Tachibana, E., Tominaga, Y., Tada, K., . . . Yagi, T. (2010). Early induction of hypothermia during cardiac arrest improves neurological outcomes in patients with out-of-hospital cardiac arrest who undergo emergency cardiopulmonary bypass and percutaneous coronary intervention. *Circ J, 74*(1), 77–85.

Negovsky, V. A. (1972). The second step in resuscitation: the treatment of the "post-resuscitation disease." *Resuscitation, 1*(1), 1–7.

Negovsky, V. A., & Gurvitch, A. M. (1995). Post-resuscitation disease: a new nosological entity—its reality and significance. *Resuscitation, 30*(1), 23–27.

Neumar, R. W., Barnhart, J. M., Berg, R. A., Chan, P. S., Geocadin, R. G., Luepker, R. V., . . . Advocacy Coordinating Committee. (2011). Implementation strategies for improving survival after out-of-hospital cardiac arrest in the United States: consensus recommendations from the 2009 American Heart Association Cardiac Arrest Survival Summit. *Circulation, 123*(24), 2898–2910.

Neumar, R. W., Nolan, J. P., Adrie, C., Aibiki, M., Berg, R. A., Bottiger, B. W., . . . Vanden Hoek, T. (2008). Post-cardiac arrest syndrome: epidemiology, pathophysiology, treatment, and prognostication—a consensus statement from the International Liaison Committee on Resuscitation (American Heart Association, Australian and New Zealand Council on Resuscitation, European Resuscitation Council, Heart and Stroke Foundation of Canada, InterAmerican Heart Foundation, Resuscitation Council of Asia, and the Resuscitation Council of Southern Africa); the American Heart Association Emergency Cardiovascular Care Committee; the Council on Cardiovascular Surgery and Anesthesia; the Council on Cardiopulmonary, Perioperative, and Critical Care; the Council on Clinical Cardiology; and the Stroke Council. *Circulation, 118*(23), 2452–2483.

Newman, D. H., Callaway, C. W., Greenwald, I. B., & Freed, J. (2004). Cerebral oximetry in out-of-hospital cardiac arrest: standard CPR rarely provides detectable hemoglobin-oxygen saturation to the frontal cortex. *Resuscitation, 63*(2), 189–194. doi: 10.1016/j.resuscitation.2004.05.003.

Nichol, G., Aufderheide, T. P., Eigel, B., Neumar, R. W., Lurie, K. G., Bufalino, V. J., . . . Outcomes Research. (2010). Regional systems of care for out-of-hospital cardiac arrest: a policy statement from the American Heart Association. *Circulation, 121*(5), 709–729.

Nichol, G., Thomas, E., Callaway, C. W., Hedges, J., Powell, J. L., Aufderheide, T. P., . . . Resuscitation Outcomes Consortium Investigators. (2008). Regional variation in out-of-hospital cardiac arrest incidence and outcome. *JAMA, 300*(12), 1423–1431.

Nishizawa, H., & Kudoh, I. (1996). Cerebral autoregulation is impaired in patients resuscitated after cardiac arrest. *Acta Anaesthesiol Scand, 40*(9), 1149–1153.

Nolan, J. P. (2011). Optimizing outcome after cardiac arrest. *Curr Opin Crit Care, 17*(5), 520–526.

Nolan, J. P., Laver, S. R., Welch, C. A., Harrison, D. A., Gupta, V., & Rowan, K. (2007). Outcome following admission to UK intensive care units after cardiac arrest: a secondary analysis of the ICNARC Case Mix Programme Database. *Anaesthesia, 62*(12), 1207–1216.

Nolan, J. P., Morley, P. T., Vanden Hoek, T. L., Hickey, R. W., Kloeck, W. G., Billi, J., . . . International Liaison Committee on Resuscitation. (2003). Therapeutic hypothermia after cardiac arrest: an advisory statement by the advanced life support task force of the International Liaison Committee on Resuscitation. *Circulation, 108*(1), 118–121.

Nolan, J. P., & Soar, J. (2011). Does the evidence support the use of mild hypothermia after cardiac arrest? Yes. *BMJ, 343,* d5830.

Nolan, J. P., & Soar, J. (2011). Mild therapeutic hypothermia after cardiac arrest: keep on chilling. *Crit Care Med, 39*(1), 206–207.

Nolan, J. P., Soar, J., Wenzel, V., & Paal, P. (2012). Cardiopulmonary resuscitation and management of cardiac arrest. *Nat Rev Cardiol, 9*(9), 499–511.

O'Neill, J. F., & Deakin, C. D. (2007). Do we hyperventilate cardiac arrest patients? *Resuscitation, 73*(1), 82–85.

O'Reilly, S. M., Grubb, N., & O'Carroll, R. E. (2004). Long-term emotional consequences of in-hospital cardiac arrest and myocardial infarction. *Br J Clin Psychol, 43*(Pt 1), 83–95.

O'Reilly, S. M., Grubb, N. R., & O'Carroll, R. E. (2003). In-hospital cardiac arrest leads to chronic memory impairment. *Resuscitation, 58*(1), 73–79.

Oddo, M. (2012). Prognostication after cardiac arrest: time to change our approach. *Resuscitation, 83*(2), 149–150.

Oddo, M., & Rossetti, A. O. (2011). Predicting neurological outcome after cardiac arrest. *Curr Opin Crit Care, 17*(3), 254–259.

Oddo, M., Schaller, M. D., Feihl, F., Ribordy, V., & Liaudet, L. (2006). From evidence to clinical practice: effective implementation of therapeutic hypothermia to improve patient outcome after cardiac arrest. *Crit Care Med, 34*(7), 1865–1873.

Odegaard, S., Olasveengen, T., Steen, P. A., & Kramer-Johansen, J. (2009). The effect of transport on quality of cardiopulmonary resuscitation in out-of-hospital cardiac arrest. *Resuscitation, 80*(8), 843–848.

Oksanen, T., Pettila, V., Hynynen, M., Varpula, T., & Intensium Consortium Study Group. (2007). Therapeutic hypothermia after cardiac arrest: implementation and outcome in Finnish intensive care units. *Acta Anaesthesiol Scand, 51*(7), 866–871.

Ong, M. E., Ornato, J. P., Edwards, D. P., Dhindsa, H. S., Best, A. M., Ines, C. S., . . . Peberdy, M. A. (2006). Use of an automated, load-distributing band chest compression device for out-of-hospital cardiac arrest resuscitation. *JAMA, 295*(22), 2629–2637.

Orban, J. C., Cattet, F., Lefrant, J. Y., Leone, M., Jaber, S., Constantin, J. M., . . . for the AzuRea Group. (2012). The practice of therapeutic hypothermia after cardiac arrest in France: A national survey. *PLoS One, 7*(9), e45284.

Ornato, J. P. (1984). "Life after life" experiences during cardiac arrest: a case report. *Nebr Med J, 69*(5), 153–155.

Ornato, J. P., & Peberdy, M. A. (2006). Prehospital and emergency department care to preserve neurologic function during and following cardiopulmonary resuscitation. *Neurol Clin, 24*(1), 23–39.

Ornato, J. P., Peberdy, M. A., Reid, R. D., Feeser, V. R., Dhindsa, H. S., & NRCPR Investigators. (2012). Impact of resuscitation system errors on survival from in-hospital cardiac arrest. *Resuscitation, 83*(1), 63–69.

Palmer, T. D., Schwartz, P. H., Taupin, P., Kaspar, B., Stein, S. A., & Gage, F. H. (2001). Cell culture: progenitor cells from human brain after death. *Nature, 411*(6833), 42–43.

Parnia, S. (2012). Cerebral oximetry: the holy grail of non-invasive cerebral perfusion monitoring in cardiac arrest or just a false dawn? *Resuscitation,* *83*(1), 11–12.

Parnia, S., Nasir, A., Shah, C., Patel, R., Mani, A., & Richman, P. (2012). A feasibility study evaluating the role of cerebral oximetry in predicting return of spontaneous circulation in cardiac arrest. *Resuscitation, 83*(8), 982–985.

Peberdy, M. A., Callaway, C. W., Neumar, R. W., Geocadin, R. G., Zimmerman, J. L., Donnino, M., . . . Kronick, S. L. (2010). Part 9: post-cardiac arrest care—2010 American Heart Association Guidelines for Cardiopulmonary Resuscitation and Emergency Cardiovascular Care. *Circulation, 122*(18 Suppl 3), S768–786.

Peberdy, M. A., Kaye, W., Ornato, J. P., Larkin, G. L., Nadkarni, V., Mancini, M. E., . . . Lane-Trultt, T. (2003). Cardiopulmonary resuscitation of adults in the hospital: a report of 14720 cardiac arrests from the National Registry of Cardiopulmonary Resuscitation. *Resuscitation, 58*(3), 297–308.

Peberdy, M. A., Ornato, J. P., Larkin, G. L., Braithwaite, R. S., Kashner, T. M., Carey, S. M., . . . National Registry of Cardiopulmonary Resuscitation Investigators. (2008). Survival from in-hospital cardiac arrest during nights and weekends. *JAMA, 299*(7), 785–792.

Perkins, G. D., Brace, S. J., Smythe, M., Ong, G., & Gates, S. (2012). Out-of-hospital cardiac arrest: recent advances in resuscitation and effects on outcome. *Heart, 98*(7), 529–535. doi: 10.1136/heartjnl-2011–300802.

Perkins, G. D., & Cooke, M. W. (2012). Variability in cardiac arrest survival: the NHS Ambulance Service Quality Indicators. *Emerg Med J, 29*(1), 3–5. doi: 10.1136/emermed-2011–200758.

Perkins, G. D., & Soar, J. (2005). In hospital cardiac arrest: missing links in the chain of survival. *Resuscitation, 66*(3), 253–255. doi: 10.1016/j.resuscitation.2005.05.010.

Perkins, G. D., Woollard, M., Cooke, M. W., Deakin, C., Horton, J., Lall, R., . . . PARAMEDIC Trial Collaborators. (2010). Prehospital randomised assessment of a mechanical compression device in cardiac arrest (PARAMEDIC) trial protocol. *Scand J Trauma Resusc Emerg Med, 18*I 58. doi: 10.1186/1757–7241–18–58.

Pilcher, J., Weatherall, M., Shirtcliffe, P., Bellomo, R., Young, P., & Beasley, R. (2012). The effect of hyperoxia following cardiac arrest: a systematic review and meta-analysis of animal trials. *Resuscitation, 83*(4), 417–422.

Pilkington, S. N., Hett, D. A., Pierce, J. M., & Smith, D. C. (1995). Auditory evoked responses and near infrared spectroscopy during cardiac arrest. *Br J Anaesth, 74*(6), 717–719.

Polderman, K. H. (2008). Hypothermia and neurological outcome after cardiac arrest: state of the art. *Eur J Anaesthesiol Suppl, 42,* 23–30.

Prodhan, P., Fiser, R. T., Dyamenahalli, U., Gossett, J., Imamura, M., Jaquiss, R. D., & Bhutta, A. T. (2009). Outcomes after extracorporeal cardiopulmonary resuscitation (ECPR) following refractory pediatric cardiac arrest in the intensive care unit. *Resuscitation, 80*(10), 1124–1129.

Putzer, G., Tiefenthaler, W., Mair, P., & Paal, P. (2012). Near-infrared spectroscopy during cardiopulmonary resuscitation of a hypothermic polytraumatised cardiac arrest patient. *Resuscitation, 83*(1), e1–2.

Redpath, C., Sambell, C., Stiell, I., Johansen, H., Williams, K., Samie, R., . . . Birnie, D. (2010). In-hospital mortality in 13,263 survivors of out-of-hospital cardiac arrest in Canada. *Am Heart J, 159*(4), 577–583, e571.

Reinikainen, M., Oksanen, T., Leppanen, P., Torppa, T., Niskanen, M., Kurola, J., & Finnish Intensive Care Consortium. (2012). Mortality in out-of-hospital cardiac arrest patients has decreased in the era of therapeutic hypothermia. *Acta Anaesthesiol Scand, 56*(1), 110–115.

Reynolds, J. C., & Lawner, B. J. (2012). Management of the post-cardiac arrest syndrome. *J Emerg Med, 42*(4), 440–449.

Reynolds, J. C., Salcido, D. D., & Menegazzi, J. J. (2010). Coronary perfusion pressure and return of spontaneous circulation after prolonged cardiac arrest. *Prehosp Emerg Care, 14*(1), 78–84. doi: 10.3109/10903120903349796.

Rivers, E. P., Rady, M. Y., Martin, G. B., Fenn, N. M., Smithline, H. A., Alexander, M. E., & Nowak, R. M. (1992). Venous hyperoxia after cardiac arrest. Characterization of a defect in systemic oxygen utilization. *Chest, 102*(6), 1787–1793.

Ro, Y. S., Shin, S. D., Song, K. J., Park, C. B., Lee, E. J., Ahn, K. O., & Cho, S. I. (2012). A comparison of outcomes of out-of-hospital cardiac arrest with non-cardiac etiology between emergency departments with low- and high-resuscitation case volume. *Resuscitation, 83*(7), 855–861.

Safar, P. (1983). Cerebral resuscitation after cardiac arrest: summaries and suggestions. *Am J Emerg Med, 1*(2), 198–214.

Safar, P. (1985). Effects of the postresuscitation syndrome on cerebral recovery from cardiac arrest. *Crit Care Med, 13*(11), 932–935.

Samaniego, E. A., Persoon, S., & Wijman, C. A. (2011). Prognosis after cardiac arrest and hypothermia: a new paradigm. *Curr Neurol Neurosci Rep, 11*(1), 111–119.

Sanchez-Lazaro, I. J., Almenar-Bonet, L., Martinez-Dolz, L., Buendia-Fuentes, F., Aguero, J., Navarro-Manchon, J., . . . Salvador-Sanz, A. (2010). Can we accept donors who have suffered a resuscitated cardiac arrest? *Transplant Proc, 42*(8), 3091–3092.

Sander, M., von Heymann, C., & Spies, C. (2006). Implementing the International Liaison Committee on Resuscitation guidelines on hypothermia after cardiac arrest: the German experience—still a long way to go? *Crit Care, 10*(2), 407.

Sanders, A. B. (2008). Progress in improving neurologically intact survival from cardiac arrest. *Ann Emerg Med, 52*(3), 253–255.

Sandroni, C., Adrie, C., Cavallaro, F., Marano, C., Monchi, M., Sanna, T., & Antonelli, M. (2010). Are patients brain-dead after successful resuscitation from cardiac arrest suitable as organ donors? A systematic review. *Resuscitation, 81*(12), 1609–1614.

Sasson, C., Rogers, M. A., Dahl, J., & Kellermann, A. L. (2010). Predictors of survival from out-of-hospital cardiac' arrest: a systematic review and meta-analysis. *Circ Cardiovasc Qual Outcomes, 3*(1), 63–81.

Schneider, A., Albertsmeier, M., Bottiger, B. W., & Teschendorf, P. (2012). [Post-resuscitation syndrome: role of inflammation after cardiac arrest]. *Anaesthesist, 61*(5), 424–436.

Scholefield, B. R., Duncan, H. P., & Morris, K. P. (2010). Survey of the use of therapeutic hypothermia post cardiac arrest. *Arch Dis Child, 95*(10), 796–799.

Scolletta, S., Taccone, F. S., Nordberg, P., Donadello, K., Vincent, J. L., & Castren, M. (2012). Intra-arrest hypothermia during cardiac arrest: a systematic review. *Crit Care, 16*(2), R41.

Shemie, S. D. (2007). Brain arrest, cardiac arrest and uncertainties in defining death. *J Pediatr (Rio J), 83*(2), 102–104.

Shemie, S. D., Langevin, S., & Farrell, C. (2010). Therapeutic hypothermia after cardiac arrest: another confounding factor in brain-death testing. *Pediatr Neurol, 42*(4), 304; author reply 304–305. doi: 10.1016/j.pediatrneurol.2010.01.011.

Soar, J., & Nolan, J. P. (2007). Mild hypothermia for post cardiac arrest syndrome. *BMJ, 335*(7618), 459–460.

Soholm, H., Wachtell, K., Nielsen, S. L., Bro-Jeppesen, J., Pedersen, F., Wanscher, M., . . . Kjaergaard, J. (2012). Tertiary centres have improved survival compared to other hospitals in the Copenhagen area after out-of-hospital cardiac arrest. *Resuscitation.* doi: 10.1016/j.resuscitation.2012.06.029.

Spaulding, C. M., Joly, L. M., Rosenberg, A., Monchi, M., Weber, S. N., Dhainaut, J. F., & Carli, P. (1997). Immediate coronary angiography in survivors of out-of-hospital cardiac arrest. *N Engl J Med, 336*(23), 1629–1633.

Stephenson, H. E., Jr., Reid, L. C., & Hinton, J. W. (1953). Some common denominators in 1200 cases of cardiac arrest. *Ann Surg, 137*(5), 731–744.

Suffoletto, B., Kristan, J., Rittenberger, J. C., Guyette, F., Hostler, D., & Callaway, C. (2012). Near-infrared spectroscopy in post-cardiac arrest patients undergoing therapeutic hypothermia. *Resuscitation, 83*(8), 986–990.

Sunde, K., Pytte, M., Jacobsen, D., Mangschau, A., Jensen, L. P., Smedsrud, C., . . . Steen, P. A. (2007). Implementation of a standardised treatment protocol for post resuscitation care after out-of-hospital cardiac arrest. *Resuscitation, 73*(1), 29–39.

Sunde, K., & Soreide, E. (2011). Therapeutic hypothermia after cardiac arrest: where are we now? *Curr Opin Crit Care, 17*(3), 247–253.

Tagami, T., Hirata, K., Takeshige, T., Matsui, J., Takinami, M., Satake, M., . . . Hirama, H. (2012). Implementation of the fifth link of the chain of survival concept for out-of-hospital cardiac arrest. *Circulation, 126*(5), 589–597.

Takasu, A., Yagi, K., Ishihara, S., & Okada, Y. (1995). Combined continuous monitoring of systemic and cerebral oxygen metabolism after cardiac arrest. *Resuscitation, 29*(3), 189–194.

Tasker, R. C. (2009). Extracorporeal cardiopulmonary resuscitation for in-hospital cardiac arrest: lessons from acute neurotoxicity. *Pediatr Crit Care Med, 10*(4), 525–527.

Testa, A., Cibinel, G. A., Portale, G., Forte, P., Giannuzzi, R., Pignataro, G., & Silveri, N. G. (2010). The proposal of an integrated ultrasonographic approach into the ALS algorithm for cardiac arrest: the PEA protocol. *Eur Rev Med Pharmacol Sci, 14*(2), 77–88.

Testori, C., Sterz, F., Behringer, W., Haugk, M., Uray, T., Zeiner, A., . . . Losert, H. (2011). Mild therapeutic hypothermia is associated with favourable outcome in patients after cardiac arrest with non-shockable rhythms. *Resuscitation, 82*(9), 1162–1167.

Testori, C., Sterz, F., Holzer, M., Losert, H., Arrich, J., Herkner, H., . . . Uray, T. (2012). The beneficial effect of mild therapeutic hypothermia depends on the time of complete circulatory standstill in patients with cardiac arrest. *Resuscitation, 83*(5), 596–601.

Thayne, R. C., Thomas, D. C., Neville, J. D., & Van Dellen, A. (2005). Use of an impedance threshold device improves short-term outcomes following out-of-hospital cardiac arrest. *Resuscitation, 67*(1), 103–108.

Tiainen, M., Poutiainen, E., Kovala, T., Takkunen, O., Happola, O., & Roine, R. O. (2007). Cognitive and neurophysiological outcome of cardiac arrest survivors treated with therapeutic hypothermia. *Stroke, 38*(8), 2303–2308.

Tirschwell, D. (2006). Optimizing neurologic prognosis after cardiac arrest. *Crit Care, 10*(6), 171.

Tomte, O., Draegni, T., Mangschau, A., Jacobsen, D., Auestad, B., & Sunde, K. (2011). A comparison of intravascular and surface cooling techniques in comatose cardiac arrest survivors. *Crit Care Med, 39*(3), 443–449.

Urban, P., Scheidegger, D., Buchmann, B., & Barth, D. (1988). Cardiac arrest and blood ionized calcium levels. *Ann Intern Med, 109*(2), 110–113.

Vaillancourt, C., Everson-Stewart, S., Christenson, J., Andrusiek, D., Powell, J., Nichol, G., . . . Resuscitation Outcomes Consortium Investigators. (2011). The impact of increased chest compression fraction on return of spontaneous circulation for out-of-hospital cardiac arrest patients not in ventricular fibrillation. *Resuscitation, 82*(12), 1501–1507. doi: 10.1016/j. resuscitation.2011.07.011.

van Alem, A. P., de Vos, R., Schmand, B., & Koster, R. W. (2004). Cognitive impairment in survivors of out-of-hospital cardiac arrest. *Am Heart J, 148*(3), 416–421.

van der Hoeven, J. G., de Koning, J., Compier, E. A., & Meinders, A. E. (1995). Early jugular bulb oxygenation monitoring in comatose patients after an out-of-hospital cardiac arrest. *Intensive Care Med, 21*(7), 567–572.

van der Wal, G., Brinkman, S., Bisschops, L. L., Hoedemaekers, C. W., van der Hoeven, J. G., de Lange, D. W., . . . Pickkers, P. (2011). Influence of mild therapeutic hypothermia after cardiac arrest on hospital mortality. *Crit Care Med, 39*(1), 84–88.

van Genderen, M. E., Lima, A., Akkerhuis, M., Bakker, J., & van Bommel, J. (2012). Persistent peripheral and microcirculatory perfusion alterations after out-of-hospital cardiac arrest are associated with poor survival. *Crit Care Med, 40*(8), 2287–2294.

Vanston, V. J., Lawhon-Triano, M., Getts, R., Prior, J., & Smego, R. A., Jr. (2010). Predictors of poor neurologic outcome in patients undergoing therapeutic hypothermia after cardiac arrest. *South Med J, 103*(4), 301–306.

Varon, J., & Marik, P. E. (2007). Steroids in cardiac arrest: not ready for prime time? *Am J Emerg Med, 25*(3), 376–377.

Wachelder, E. M., Moulaert, V. R., van Heugten, C., Verbunt, J. A., Bekkers, S. C., & Wade, D. T. (2009). Life after survival: long-term daily functioning and quality of life after an out-of-hospital cardiac arrest. *Resuscitation, 80*(5), 517–522.

Wagner, H., Terkelsen, C. J., Friberg, H., Harnek, J., Kern, K., Lassen, J. F., & Olivecrona, G. K. (2010). Cardiac arrest in the catheterisation laboratory:

a 5-year experience of using mechanical chest compressions to facilitate PCI during prolonged resuscitation efforts. *Resuscitation, 81*(4), 383–387.

Walters, E. L., Morawski, K., Dorotta, I., Ramsingh, D., Lumen, K., Bland, D., ... Nguyen, H. B. (2011). Implementation of a post-cardiac arrest care bundle including therapeutic hypothermia and hemodynamic optimization in comatose patients with return of spontaneous circulation after out-of-hospital cardiac arrest: a feasibility study. *Shock, 35*(4), 360–366.

Walters, J. H., Morley, P. T., & Nolan, J. P. (2011). The role of hypothermia in post-cardiac arrest patients with return of spontaneous circulation: a systematic review. *Resuscitation, 82*(5), 508–516.

Wang, H. E., Devlin, S. M., Sears, G. K., Vaillancourt, C., Morrison, L. J., Weisfeldt, M., ... Resuscitation Outcomes Consortium Investigators. (2012). Regional variations in early and late survival after out-of-hospital cardiac arrest. *Resuscitation, 83*(11), 1343–1348.

Wang, H. E., Szydlo, D., Stouffer, J. A., Lin, S., Carlson, J. N., Vaillancourt, C., ... Resuscitation Outcomes Consortium Investigators. (2012). Endotracheal intubation versus supraglottic airway insertion in out-of-hospital cardiac arrest. *Resuscitation, 83*(9), 1061–1066.

Wang, H. E., Thomas, J. J., James, D., Barlotta, K., Fellman, A., Viles, A., ... Lai, K. R. (2011). Post-cardiac arrest therapeutic hypothermia: overcoming the barrier of workplace culture and other implementation lessons. *Jt Comm J Qual Patient Saf, 37*(9), 425–432.

White, N. J., Leong, B. S., Brueckner, J., Martin, E. J., Brophy, D. F., Peberdy, M. A., ... Ward, K. R. (2011). Coagulopathy during cardiac arrest and resuscitation in a swine model of electrically induced ventricular fibrillation. *Resuscitation, 82*(7), 925–931.

Wik, L., Kramer-Johansen, J., Myklebust, H., Sorebo, H., Svensson, L., Fellows, B., & Steen, P. A. (2005). Quality of cardiopulmonary resuscitation during out-of-hospital cardiac arrest. *JAMA, 293*(3), 299–304.

Won, S. J., Kim, D.Y., & Gwag, B. J. (2002). Cellular and molecular pathways of ischemic neuronal death. *J Biochem Mol Biol, 35*(1), 67–86.

Yannopoulos, D., Matsuura, T., McKnite, S., Goodman, N., Idris, A., Tang, W., ... Lurie, K. G. (2010). No assisted ventilation cardiopulmonary resuscitation and 24-hour neurological outcomes in a porcine model of cardiac arrest. *Crit Care Med, 38*(1), 254–260. doi: 10.1097/CCM.0b013e3181b42f6c.

Yannopoulos, D., Matsuura, T., Schultz, J., Rudser, K., Halperin, H. R., & Lurie, K. G. (2011). Sodium nitroprusside enhanced cardiopulmonary resuscitation improves survival with good neurological function in a porcine model of prolonged cardiac arrest. *Crit Care Med, 39*(6), 1269–1274.

Yannopoulos, D., McKnite, S., Aufderheide, T. P., Sigurdsson, G., Pirrallo, R. G., Benditt, D., & Lurie, K. G. (2005). Effects of incomplete chest wall decompression during cardiopulmonary resuscitation on coronary and cerebral perfusion pressures in a porcine model of cardiac arrest. *Resuscitation, 64*(3), 363–372.

Yannopoulos, D., Nadkarni, V. M., McKnite, S. H., Rao, A., Kruger, K., Metzger, A., . . . Lurie, K. G. (2005). Intrathoracic pressure regulator during continuous-chest-compression advanced cardiac resuscitation improves vital organ perfusion pressures in a porcine model of cardiac arrest. *Circulation, 112*(6), 803–811.

Yannopoulos, D., Sigurdsson, G., McKnite, S., Benditt, D., & Lurie, K. G. (2004). Reducing ventilation frequency combined with an inspiratory impedance device improves CPR efficiency in swine model of cardiac arrest. *Resuscitation, 61*(1), 75–82.

Young, W. L., & Ornstein, E. (1985). Compressed spectral array EEG monitoring during cardiac arrest and resuscitation. *Anesthesiology, 62*(4), 535–538.

Near-Death Experiences/Actual-Death Experiences

Appleton, R. E. (1993). Reflex anoxic seizures. *BMJ, 307*(6898), 214–15.

Bates, B. C., & Stanley, A. (1985). The epidemiology and differential diagnosis of near-death experience. *Am J Orthopsychiatry, 55*(4), 542–549.

Blackmore, S. J. (1996). Near death experiences. *J Royal Soc Med, 89*(2), 73–76.

Blackmore, S. J., & Troscianko, T. (1988). The physiology of the tunnel. *J Near Death Stud, 8*(1), 15–28.

Blanke, O. (2004). Out of body experiences and their neural basis. *BMJ, 329*(7480), 1414–1415.

Blanke, O., Ortigue, S., Landis, T., & Seeck, M. (2002). Stimulating illusory own-body perceptions. *Nature, 419*(6904), 269–270.

Bonta, I. L. (2004). Schizophrenia, dissociative anaesthesia and near-death experience: three events meeting at the NMDA receptor. *Med Hypotheses, 62*(1), 23–28.

Britton, W. B., & Bootzin, R. R. (2004). Near-death experiences and the temporal lobe. *Psychol Sci, 15*(4), 254–258.

Carr, D. (1982). Pathophysiology of stress-induced limbic lobe dysfunction: a hypothesis for NDEs. *J Near Death Stud, 2*(2), 75–89.

Carr, D. B. (1981). Endorphins at the approach of death. *Lancet, 1*(8216), 390.

Dougherty, C. M. (1990). The near-death experience as a major life transition. *Holist Nurs Pract, 4*(3), 84–90.

Duffy, N., & Olson, M. (2007). Supporting a patient after a near-death experience. *Nursing, 37*(4), 46–48. doi: 10.1097/01.NURSE .0000266041.11793.9d.

Feng, Z. (1992). A research on near death experiences of survivors in big earthquake of Tangshan, 1976. *Chung Hua Shen Ching Ching Shen KO Tsa Chih, 25*(Aug.), 222–225, 253–254.

Fenwick, P., & Fenwick, E. (1995). *The Truth in the Light*. London: Hodder Headline.

French, C. C. (2001). Dying to know the truth: visions of a dying brain, or false memories? *Lancet, 358*(9298), 2010–2011.

Gallup, G. (1982). *Adventures in Immortality: A Look Beyond the Threshold of Death*. New York: McGraw-Hill.

Gordon, B. D. (1989). Near-death experience. *Lancet, 2*(8677), 1452.

Greyson, B. (1983). The near-death experience scale: construction, reliability, and validity. *J Nerv Ment Dis, 171*(6), 369–375.

Greyson, B. (1993). Varieties of near-death experience. *Psychiatry, 56*(4), 390–399.

Greyson, B. (1997). The near-death experience as a focus of clinical attention. *J Nerv Ment Dis, 185*(5), 327–334.

Greyson, B. (2000). Dissociation in people who have near-death experiences: out of their bodies or out of their minds? *Lancet, 355*(9202), 460–463.

Greyson, B. (2003). Incidence and correlates of near-death experiences in a cardiac care unit. *Gen Hosp Psychiatry, 25*(4), 269–276.

Greyson, B. (2007). Consistency of near-death experience accounts over two decades: are reports embellished over time? *Resuscitation, 73*(3), 407–411. doi: 10.1016/j.resuscitation.2006.10.013.

Greyson, B. (2008). The near-death experience. *Altern Ther Health Med, 14*(3), 14; author reply 14–15.

Herzog, D. B., & Herrin, J. T. (1985). Near-death experiences in the very young. *Crit Care Med, 13*(12), 1074–1075.

Jansen, K. (1989). Near death experience and the NMDA receptor. *BMJ, 298*(6689), 1708.

Jansen, K. L. (1989). The near-death experience. *Br J Psychiatry, 154*, 883–884.

Jansen, K. L. (1990). Neuroscience and the near-death experience: roles for the NMSA-PCP receptor, the sigma receptor and the endopsychosins. *Med Hypotheses, 31*(1), 25–29.

Jansen, K. L. (1991). Transcendental explanations and near-death experience. *Lancet, 337*(8735), 244.

Judson, I. R., & Wiltshaw, E. (1983). A near-death experience. *Lancet, 2*(8349), 561–562.

Kellehear, A. (1990). The near-death experience as status passage. *Soc Sci Med, 31*(8), 933–939.

Kellehear, A. (1993). Culture, biology, and the near-death experience: a reappraisal. *J Nerv Ment Dis, 181*(3), 148–156.

Kelly, E. W. (2001). Near-death experiences with reports of meeting deceased people. *Death Stud, 25*(3), 229–49.

Klemenc-Ketis, Z. (2011). Life changes in patients after out-of-hospital cardiac arrest: the effect of near-death experiences. *Int J Behav Med.* doi: 10.1007/s12529-011-9209-y.

Klemenc-Ketis, Z., Kersnik, J., & Grmec, S. (2010). The effect of carbon dioxide on near-death experiences in out-of-hospital cardiac arrest survivors: a prospective observational study. *Crit Care, 14*(2), R56.

Lempert, T. (1994). Syncope and near death experience. *Lancet, 344*(8925), 829–830.

Lenggenhager, B., Tadi, T., Metzinger, T., & Blanke, O. (2007). Video ergo sum: manipulating bodily self-consciousness. *Science, 317*(5841), 1096–1099.

Martens, P. R. (1994). Near-death-experiences in out-of-hospital cardiac arrest survivors: meaningful phenomena or just fantasy of death? *Resuscitation, 27*(2), 171–175.

Meduna, L. J. (1958). *Carbon Dioxide Therapy: A Neurophysiological Treatment of Nervous Disorders* (2nd ed.). Springfield, IL: Charles C. Thomas.

Mobbs, D., & Watt, C. (2011). There is nothing paranormal about near-death experiences: how neuroscience can explain seeing bright lights, meeting the dead, or being convinced you are one of them. *Trends Cogn Sci, 15*(10), 447–449.

Moody, R. A. (1975). *Life After Life.* New York: Bantam Press.

Morse, M., Castillo, P., Venecia, D., Milstein, J., & Tyler, D. C. (1986). Childhood near death experiences. *Am J Dis Child, 140*(11), 1110–1114.

Morse, M., Venecia, D., & Milstein, J. (1989). Near-death experiences: a neurophysiologic explanatory model. *J Near Death Stud, 8*(1), 45–53.

Morse, M. L. (1994). Near death experiences of children. *J Ped Onc Nurs,* *11,* 139–44.

Nelson, K. R., Mattingly, M., Lee, S. A., & Schmitt, F. A. (2006). Does the arousal system contribute to near death experience? *Neurology, 66*(7), 1003–1009.

Noyes, R., & Kletti, R. (1976). Depersonalisation in the face of life threatening danger: a description. *Psychiatry, 39,* 251–259.

Orne, R. M. (1995). The meaning of survival: the early aftermath of a near-death experience. *Res Nurs Health, 18*(3), 239–247.

Osis, K., & Haraldsson, E. (1977). *At the Hour of Death.* New York: Avon Books.

Owens, J. E., Cook, E. W., & Stevenson, I. (1990). Features of "near-death experience" in relation to whether or not patients were near death. *Lancet, 336*(8724), 1175–1177.

Owens, J. E., Cook, E. W., & Stevenson, I. (1991). Near-death experience. *Lancet, 337*(8750), 1167–1168.

Parnia, S. (2007). Do reports of consciousness during cardiac arrest hold the key to discovering the nature of consciousness? *Med Hypotheses, 69*(4), 933–937. doi: 10.1016/j.mehy.2007.01.076.

Parnia, S., & Fenwick, P. (2002). Near death experiences in cardiac arrest: visions of a dying brain or visions of a new science of consciousness? *Resuscitation, 52*(1), 5–11.

Parnia, S., Waller, D., Yeates, R., & Fenwick, P. (2001). A qualitative and quantitative study of the incidence, features and aetiology of near death experiences in cardiac arrest survivors. *Resuscitation, 48*(2), 149–156.

Pascricha, S., & Stevenson, I. (1986). Near death experiences in India. *J Nerv Ment Dis, 55*(4), 542–549.

Sabom, M. B. (1980). The near-death experience. *JAMA, 244*(1), 29–30.

Schwaninger, J. (2002). A prospective analysis of near death experiences in cardiac arrest patients. *J Near Death Stud,* 20(4).

Serdahely, W., Drenk, A., & Serdahely, J. J. (1988). What carers need to understand about the near-death experience. *Geriatr Nurs, 9*(4), 238–241.

Serdahely, W. J. (1990). Pediatric near death experiences. *J Near Death Stud, 9*(1), 33–39.

Serdahely, W. J. (1992). The near-death experience and caregivers: helping and being helped. *Caring, 11*(1), 8–11.

Sommers, M. S. (1994). The near-death experience following multiple trauma. *Crit Care Nurse, 14*(2), 62–66.

Sotelo, J., Perez, R., Guevara, P., & Fernandez, A. (1995). Changes in brain, plasma and cerebrospinal fluid contents of B-endorphin in dogs at the moment of death. *Neurol Res, 17*(June), 223.

Van Lommel, P. (2011). Near-death experiences: the experience of the self as real and not as an illusion. *Ann NY Acad Sci, 1234,* 19–28.

Van Lommel, P., van Wees, R., Meyers, V., & Elfferich, I. (2001). Near-death experience in survivors of cardiac arrest: a prospective study in the Netherlands. *Lancet, 358*(9298), 2039–2045.

van Tellingen, C. (2008). Heaven can wait—or down to earth in real time: near-death experience revisited. *Neth Heart J, 16*(10), 359–362.

Walker, F. O. (1989). A nowhere near-death experience: heavenly choirs interrupt myelography. *JAMA, 261*(22), 3245–3246.

Williams, B. (1993). Near-death experience denied. *CMAJ, 148*(3), 376.

Yamamura, H. (1998). [Implication of near-death experience for the elderly in terminal care]. *Nihon Ronen Igakkai Zasshi, 35*(2), 103–115.

Brain, Consciousness, Soul

Baig, M. N., Chishty, F., Immesoete, P., & Karas, C. S. (2007). The Eastern heart and Galen's ventricle: a historical review of the purpose of the brain. *Neurosurg Focus, 23*(1), E3.

Beck, F., & Eccles, J. C. (1992). Quantum aspects of brain activity and the role of consciousness. *Proc Natl Acad Sci U S A, 89*(23), 11357–11361.

Bennett, M. R. (2007). Development of the concept of mind. *Aust N Z J Psychiatry, 41*(12), 943–956.

Blackmore, S. (2003). *Consciousness: An Introduction.* London: Hodder and Stoughton.

Chalmers, D. J. (1997). The puzzle of conscious experience, mysteries of the mind. *Sci Am* (special issue), 30–37.

Crick, F., & Koch, C. (1998). Consciousness and neuroscience. *Cereb Cortex, 8*(2), 97–107.

Crivellato, E., & Ribatti, D. (2007). Soul, mind, brain: Greek philosophy and the birth of neuroscience. *Brain Res Bull, 71*(4), 327–336.

Demertzi, A., Liew, C., Ledoux, D., Bruno, M. A., Sharpe, M., Laureys, S., & Zeman, A. (2009). Dualism persists in the science of mind. *Ann NY Acad Sci, 1157,* 1–9.

Dennett, D. (2001). Are we explaining consciousness yet? *Cognition, 79*(1–2), 221–237.

Does neuroscience threaten human values? (1998). *Nat Neurosci, 1*(Nov.), 535–536.

Dolan, B. (2007). Soul searching: a brief history of the mind/body debate in the neurosciences. *Neurosurg Focus, 23*(1), E2.

Eccles, J. C. (1992). Evolution of consciousness. *Proc Natl Acad Sci U S A, 89*(16), 7320–7324.

Elahi, B. (1999). *Spirituality Is a Science.* New York: Cornwall Books, study 2.

Elahi, B. (2001). *Medicine of the Soul.* New York: Cornwall Books, studies 4 and 6.

Elahi, B. (2005). *The Path of Perfection.* New York: Paraview, chapters 5–8.

Elahi, B. (2011). Video of a lecture delivered at the University of Sorbonne, Paris, March 2011. www.e-ostadelahi.com.

Fenwick, P. (2000). Current methods of investigation in neuroscience. In M. Velmans (ed.), *Investigating Phenomenal Consciousness.* Amsterdam: John Benjamins.

Flohr, H. (1995). An information processing theory of anaesthesia. *Neuropsychologia, 33*(9), 1169–1180.

Flohr, H., Glade, U., & Motzko, D. (1998). The role of the NMDA synapse in general anesthesia. *Toxicol Lett, 100–101,* 23–29.

Frackowiak, R. (1997). *Human Brain Function.* London: Academic Press.

Freeman, W. (1999). Consciousness, intentionality and causality. *J Consciousness Stud, 6*(11–12), 143–172.

Greenfield, S. (2002). Mind, brain and consciousness. *Br J Psychiatry, 181*(Aug.), 91–93.

Hameroff, S., Nip, A., Porter, M., & Tuszynski, J. (2002). Conduction pathways in microtubules, biological quantum computation, and consciousness. *Biosystems, 64*(1–3), 149–168.

Henslin, J. (2007). *Down to Earth Sociology* (14th ed.). New York: Free Press, 277–287.

Mysteries of the mind. (1997). *Sci Am* (special issue).

Penrose, R. (1994). *Shadows of the Mind.* Oxford: Oxford University Press.

Penrose, R. (2001). Consciousness, the brain, and spacetime geometry: an addendum—some new developments on the Orch OR model for consciousness. *Ann NY Acad Sci, 929*(Apr.), 105–110.

Rees, G., Kreiman, G., & Koch, C. (2002). Neural correlates of consciousness in humans. *Nat Rev Neurosci, 3*(4), 261–270.

Santoro, G., Wood, M. D., Merlo, L., Anastasi, G. P., Tomasello, F., & Germano, A. (2009). The anatomic location of the soul from the heart, through the brain, to the whole body, and beyond: a journey through Western history, science, and philosophy. *Neurosurgery, 65*(4), 633–643; discussion 643.

Searle, J. (1998). Do we understand consciousness? *J Consciousness Stud, 5*(5–6), 718–733.

Tononi, G., & Edelman, G. M. (1998). Consciousness and complexity. *Science, 282*(5395), 1846–1851.

Brain Death, Minimally Conscious State, and Vegetative State

Boly, M., Coleman, M. R., Davis, M. H., Hampshire, A., Bor, D., Moonen, G., . . . Owen, A. M. (2007). When thoughts become action: an fMRI paradigm to study volitional brain activity in non-communicative brain injured patients. *Neuroimage, 36*(3), 979–992.

Brown, E. N., Lydic, R., & Schiff, N. D. (2010). General anesthesia, sleep, and coma. *N Engl J Med, 363*(27), 2638–2650.

Burkle, C. M., Schipper, A. M., & Wijdicks, E. F. (2011). Brain death and the courts. *Neurology, 76*(9), 837–841.

Chatelle, C., Chennu, S., Noirhomme, Q., Cruse, D., Owen, A. M., & Laureys, S. (2012). Brain-computer interfacing in disorders of consciousness. *Brain Inj, 26*(12), 1510–1522.

Christoff, K., & Owen, A. M. (2006). Improving reverse neuroimaging inference: cognitive domain versus cognitive complexity. *Trends Cogn Sci, 10*(8), 352–353.

Coleman, M. R., Bekinschtein, T., Monti, M. M., Owen, A. M., & Pickard, J. D. (2009). A multimodal approach to the assessment of patients with disorders of consciousness. *Prog Brain Res, 177,* 231–248.

Coleman, M. R., Davis, M. H., Rodd, J. M., Robson, T., Ali, A., Owen, A. M., & Pickard, J. D. (2009). Towards the routine use of brain imaging to aid the clinical diagnosis of disorders of consciousness. *Brain, 132*(Pt 9), 2541–2552.

Coleman, M. R., & Owen, A. M. (2008). Functional neuroimaging of disorders of consciousness. *Int Anesthesiol Clin, 46*(3), 147–157.

Coleman, M. R., Rodd, J. M., Davis, M. H., Johnsrude, I. S., Menon, D. K., Pickard, J. D., & Owen, A. M. (2007). Do vegetative patients retain aspects of language comprehension? Evidence from fMRI. *Brain, 130*(Pt 10), 2494–2507.

Cruse, D., Chennu, S., Chatelle, C., Bekinschtein, T. A., Fernandez-Espejo, D., Pickard, J. D., . . . Owen, A. M. (2011). Bedside detection of awareness in the vegetative state: a cohort study. *Lancet, 378*(9809), 2088–2094.

Cruse, D., Chennu, S., Chatelle, C., Fernandez-Espejo, D., Bekinschtein, T. A., Pickard, J. D., . . . Owen, A. M. (2012). Relationship between etiology and covert cognition in the minimally conscious state. *Neurology, 78*(11), 816–822.

Cruse, D., & Owen, A. M. (2010). Consciousness revealed: new insights into the vegetative and minimally conscious states. *Curr Opin Neurol, 23*(6), 656–660.

A definition of irreversible coma: report of the Ad Hoc Committee of the Harvard Medical School to Examine the Definition of Brain Death. (1968). *JAMA, 205*(6), 337–340.

Diagnosis of brain death: statement issued by the honorary secretary of the Conference of Medical Royal Colleges and Their Faculties in the United Kingdom on 11 October 1976. (1976). *BMJ, 2*(6045), 1187–1188.

Egea-Guerrero, J. J., Revuelto-Rey, J., Latronico, N., Rasulo, F. A., & Wijdicks, E. F. (2011). The case against confirmatory tests for determining brain death in adults. *Neurology, 76*(5), 489; author reply 489–490.

Fallon, S. J., Williams-Gray, C. H., Barker, R. A., Owen, A. M., & Hampshire, A. (2012). Prefrontal dopamine levels determine the balance between cognitive stability and flexibility. *Cereb Cortex*. doi: 10.1093/cercor/bhs025.

Fernandez-Espejo, D., Bekinschtein, T., Monti, M. M., Pickard, J. D., Junque, C., Coleman, M. R., & Owen, A. M. (2011). Diffusion weighted imaging distinguishes the vegetative state from the minimally conscious state. *Neuroimage, 54*(1), 103–112.

Fernandez-Espejo, D., Soddu, A., Cruse, D., Palacios, E. M., Junque, C., Vanhaudenhuyse, A., . . . Owen, A. M. (2012). A role for the default mode network in the bases of disorders of consciousness. *Ann Neurol, 72*(3), 335–343.

Fins, J. J., & Schiff, N. D. (2006). Shades of gray: new insights into the vegetative state. *Hastings Cent Rep, 36*(6), 8.

Fins, J. J., Schiff, N. D., & Foley, K. M. (2007). Late recovery from the minimally conscious state: ethical and policy implications. *Neurology, 68*(4), 304–307.

Fugate, J. E., Rabinstein, A. A., & Wijdicks, E. F. (2011). Blood pressure patterns after brain death. *Neurology, 77*(4), 399–401.

Fugate, J. E., Wijdicks, E. F., Mandrekar, J., Claassen, D. O., Manno, E. M., White, R. D., . . . Rabinstein, A. A. (2010). Predictors of neurologic outcome in hypothermia after cardiac arrest. *Ann Neurol, 68*(6), 907–914.

Fugate, J. E., Wijdicks, E. F., White, R. D., & Rabinstein, A. A. (2011). Does therapeutic hypothermia affect time to awakening in cardiac arrest survivors? *Neurology, 77*(14), 1346–1350.

Giacino, J., Fins, J. J., Machado, A., & Schiff, N. D. (2012). Central thalamic deep brain stimulation to promote recovery from chronic posttraumatic minimally conscious state: challenges and opportunities. *Neuromodulation, 15*(4), 339–349.

Giacino, J. T., Hirsch, J., Schiff, N., & Laureys, S. (2006). Functional neuroimaging applications for assessment and rehabilitation planning in patients with disorders of consciousness. *Arch Phys Med Rehabil, 87*(12 Suppl 2), S67–76.

Giacino, J. T., Schnakers, C., Rodriguez-Moreno, D., Kalmar, K., Schiff, N., & Hirsch, J. (2009). Behavioral assessment in patients with disorders of consciousness: gold standard or fool's gold? *Prog Brain Res, 177,* 33–48.

Goldfine, A. M., & Schiff, N. D. (2011). Consciousness: its neurobiology and the major classes of impairment. *Neurol Clin, 29*(4), 723–737.

Goldfine, A. M., & Schiff, N. D. (2011). What is the role of brain mechanisms underlying arousal in recovery of motor function after structural brain injuries? *Curr Opin Neurol, 24*(6), 564–569.

Goldfine, A. M., Victor, J. D., Conte, M. M., Bardin, J. C., & Schiff, N. D. (2011). Determination of awareness in patients with severe brain injury using EEG power spectral analysis. *Clin Neurophysiol, 122*(11), 2157–2168.

Goudreau, J. L., Wijdicks, E. F., & Emery, S. F. (2000). Complications during apnea testing in the determination of brain death: predisposing factors. *Neurology, 55*(7), 1045–1048.

Greer, D. M., Varelas, P. N., Haque, S., & Wijdicks, E. F. (2008). Variability of brain death determination guidelines in leading US neurologic institutions. *Neurology, 70*(4), 284–289.

Kobylarz, E. J., & Schiff, N. D. (2004). Functional imaging of severely brain-injured patients: progress, challenges, and limitations. *Arch Neurol, 61*(9), 1357–1360.

Kobylarz, E. J., & Schiff, N. D. (2005). Neurophysiological correlates of persistent vegetative and minimally conscious states. *Neuropsychol Rehabil, 15*(3–4), 323–332.

Laureys, S., Owen, A. M., & Schiff, N. D. (2004). Brain function in coma, vegetative state, and related disorders. *Lancet Neurol, 3*(9), 537–546.

Laureys, S., Owen, A., & Schiff, N. (2009). Coma science: clinical and ethical implications. Preface. *Prog Brain Res, 177,* xiii–xiv.

Laureys, S., & Schiff, N. D. (2012). Coma and consciousness: paradigms (re)framed by neuroimaging. *Neuroimage, 61*(2), 478–491.

Leuchter, B., Pedley, T. A., Lisanby, S. H., Mayberg, H. S., & Schiff, N. D. (2012). Brain stimulation in neurology and psychiatry: perspectives on an evolving field. *Ann NY Acad Sci, 1265,* vii–x.

Liu, A. A., Voss, H. U., Dyke, J. P., Heier, L. A., & Schiff, N. D. (2011). Arterial spin labeling and altered cerebral blood flow patterns in the minimally conscious state. *Neurology, 77*(16), 1518–1523.

Lustbader, D., O'Hara, D., Wijdicks, E. F., MacLean, L., Tajik, W., Ying, A., . . . Goldstein, M. (2011). Second brain death examination may negatively affect organ donation. *Neurology, 76*(2), 119–124.

Mittal, M. K., Arteaga, G. M., & Wijdicks, E. F. (2012). Thumbs up sign in brain death. *Neurocrit Care, 17*(2), 265–267.

Mollaret, P., & Goulon, M. (1959). Le coma dépassé (mémoire préliminaire). *Rev Neurol (Paris), 101,* 3–5.

Monti, M. M., Coleman, M. R., & Owen, A. M. (2009). Executive functions in the absence of behavior: functional imaging of the minimally conscious state. *Prog Brain Res, 177,* 249–260.

Monti, M. M., Coleman, M. R., & Owen, A. M. (2009). Neuroimaging and the vegetative state: resolving the behavioral assessment dilemma? *Ann N Y Acad Sci, 1157,* 81–89.

Monti, M. M., Laureys, S., & Owen, A. M. (2010). The vegetative state. *BMJ, 341.* doi: 10.1136/bmj.c3765.

Monti, M. M., Pickard, J. D., & Owen, A. M. (2012). Visual cognition in disorders of consciousness: from V1 to top-down attention. *Hum Brain Mapp.* doi: 10.1002/hbm.21507.

Monti, M. M., Vanhaudenhuyse, A., Coleman, M. R., Boly, M., Pickard, J. D., Tshibanda, L., . . . Laureys, S. (2010). Willful modulation of brain activity in disorders of consciousness. *N Engl J Med, 362*(7), 579–589.

Muralidharan, R., Mateen, F. J., Shinohara, R. T., Schears, G. J., & Wijdicks, E. F. (2011). The challenges with brain death determination in adult patients on extracorporeal membrane oxygenation. *Neurocrit Care, 14*(3), 423–426.

Naci, L., Monti, M. M., Cruse, D., Kubler, A., Sorger, B., Goebel, R., . . . Owen, A. M. (2012). Brain-computer interfaces for communication with nonresponsive patients. *Ann Neurol, 72*(3), 312–323.

Owen, A. M. (2008). Disorders of consciousness. *Ann N Y Acad Sci, 1124,* 225–238.

Owen, A. M. (2012). Detecting consciousness: a unique role for neuroimaging. *Annu Rev Psychol.* [epub ahead of print]

Owen, A. M., & Coleman, M. R. (2007). Functional MRI in disorders of consciousness: advantages and limitations. *Curr Opin Neurol, 20*(6), 632–637.

Owen, A. M., & Coleman, M. R. (2008). Detecting awareness in the vegetative state. *Ann NY Acad Sci, 1129,* 130–138.

Owen, A. M., & Coleman, M. R. (2008). Functional neuroimaging of the vegetative state. *Nat Rev Neurosci, 9*(3), 235–243.

Owen, A. M., & Coleman, M. R. (2008). Using neuroimaging to detect awareness in disorders of consciousness. *Funct Neurol, 23*(4), 189–194.

Owen, A. M., Coleman, M. R., Boly, M., Davis, M. H., Laureys, S., & Pickard, J. D. (2006). Detecting awareness in the vegetative state. *Science, 313*(5792), 1402.

Owen, A. M., Coleman, M. R., Boly, M., Davis, M. H., Laureys, S., & Pickard, J. D. (2007). Using functional magnetic resonance imaging to detect covert awareness in the vegetative state. *Arch Neurol, 64*(8), 1098–1102.

Owen, A. M., Schiff, N. D., & Laureys, S. (2009). A new era of coma and consciousness science. *Prog Brain Res, 177,* 399–411.

Plum, F., Schiff, N., Ribary, U., & Llinas, R. (1998). Coordinated expression in chronically unconscious persons. *Philos Trans R Soc Lond B Biol Sci, 353*(1377), 1929–1933.

President's Commission for the Study of Ethical Problems in Medicine and Biomedical and Behavioral Research. (1981). *Defining Death: A Report on the Medical, Legal and Ethical Issues in the Determination of Death.* Washington, DC: Government Printing Office.

Quinkert, A. W., Schiff, N. D., & Pfaff, D. W. (2010). Temporal patterning of pulses during deep brain stimulation affects central nervous system arousal. *Behav Brain Res, 214*(2), 377–385.

Quinkert, A. W., Vimal, V., Weil, Z. M., Reeke, G. N., Schiff, N. D., Banavar, J. R., & Pfaff, D. W. (2011). Quantitative descriptions of generalized arousal, an elementary function of the vertebrate brain. *Proc Natl Acad Sci U S A, 108*(Suppl 3), 15617–15623.

Rodriguez-Moreno, D., Schiff, N. D., Giacino, J., Kalmar, K., & Hirsch, J. (2010). A network approach to assessing cognition in disorders of consciousness. *Neurology, 75*(21), 1871–1878.

Schiff, N. D. (2005). Modeling the minimally conscious state: measurements of brain function and therapeutic possibilities. *Prog Brain Res, 150,* 473–493.

Schiff, N. D. (2006). Multimodal neuroimaging approaches to disorders of consciousness. *J Head Trauma Rehabil, 21*(5), 388–397.

Schiff, N. D. (2008). Central thalamic contributions to arousal regulation and neurological disorders of consciousness. *Ann N Y Acad Sci, 1129,* 105–118.

Schiff, N. D. (2009). Central thalamic deep-brain stimulation in the severely injured brain: rationale and proposed mechanisms of action. *Ann N Y Acad Sci, 1157,* 101–116.

Schiff, N. D. (2010). Recovery of consciousness after brain injury: a meso-circuit hypothesis. *Trends Neurosci, 33*(1), 1–9.

Schiff, N. D. (2010). Recovery of consciousness after severe brain injury: the role of arousal regulation mechanisms and some speculation on the heart-brain interface. *Cleve Clin J Med, 77*(Suppl 3), S27–33.

Schiff, N. D. (2012). Moving toward a generalizable application of central thalamic deep brain stimulation for support of forebrain arousal regulation in the severely injured brain. *Ann NY Acad Sci, 1265,* 56–68.

Schiff, N. D. (2012). Posterior medial corticothalamic connectivity and consciousness. *Ann Neurol, 72*(3), 305–306.

Schiff, N. D., Giacino, J. T., & Fins, J. J. (2009). Deep brain stimulation, neuroethics, and the minimally conscious state: moving beyond proof of principle. *Arch Neurol, 66*(6), 697–702.

Schiff, N. D., Giacino, J. T., Kalmar, K., Victor, J. D., Baker, K., Gerber, M., . . . Rezai, A. R. (2007). Behavioural improvements with thalamic stimulation after severe traumatic brain injury. *Nature, 448*(7153), 600–603.

Schiff, N. D., & Laureys, S. (2009). Disorders of consciousness: preface. *Ann NY Acad Sci, 1157,* ix–xi.

Schiff, N. D., & Plum, F. (2000). The role of arousal and "gating" systems in the neurology of impaired consciousness. *J Clin Neurophysiol, 17*(5), 438–452.

Schiff, N. D., & Posner, J. B. (2007). Another "Awakenings." *Ann Neurol, 62*(1), 5–7.

Schiff, N. D., Ribary, U., Moreno, D. R., Beattie, B., Kronberg, E., Blasberg, R., . . . Plum, F. (2002). Residual cerebral activity and behavioural fragments can remain in the persistently vegetative brain. *Brain, 125*(Pt 6), 1210–1234.

Schiff, N. D., Ribary, U., Plum, F., & Llinas, R. (1999). Words without mind. *J Cogn Neurosci, 11*(6), 650–656.

Schiff, N. D., Rodriguez-Moreno, D., Kamal, A., Kim, K. H., Giacino, J. T., Plum, F., & Hirsch, J. (2005). FMRI reveals large-scale network activation in minimally conscious patients. *Neurology, 64*(3), 514–523.

Shah, S. A., Baker, J. L., Ryou, J. W., Purpura, K. P., & Schiff, N. D. (2009). Modulation of arousal regulation with central thalamic deep brain stimulation. *Conf Proc IEEE Eng Med Biol Soc,* 3314–3317. doi: 10.1109/IEMBS.2009.5333751.

Shah, S. A., & Schiff, N. D. (2010). Central thalamic deep brain stimulation for cognitive neuromodulation—a review of proposed mechanisms and investigational studies. *Eur J Neurosci, 32*(7), 1135–1144. doi: 10.1111/j.1460–9568.2010.07420.x.

Shirvalkar, P., Seth, M., Schiff, N. D., & Herrera, D. G. (2006). Cognitive enhancement with central thalamic electrical stimulation. *Proc Natl Acad Sci U S A, 103*(45), 17007–17012.

Staunton, H. (2008). Arousal by stimulation of deep-brain nuclei. *Nature, 452*(7183), E1; discussion E1–2.

Voss, H. U., & Schiff, N. D. (2009). MRI of neuronal network structure, function, and plasticity. *Prog Brain Res, 175,* 483–496.

Wijdicks, E. F. (1995). Determining brain death in adults. *Neurology, 45*(5), 1003–1011.

Wijdicks, E. F. (2001). The diagnosis of brain death. *N Engl J Med, 344*(16), 1215–1221.

Wijdicks, E. F. (2001). Topsy turvydom in brain death determination. *Transplantation, 72*(2), 355.

Wijdicks, E. F. (2002). Brain death worldwide: accepted fact but no global consensus in diagnostic criteria. *Neurology, 58*(1), 20–25.

Wijdicks, E. F. (2003). The neurologist and Harvard criteria for brain death. *Neurology, 61*(7), 970–976.

Wijdicks, E. F. (2006). The clinical criteria of brain death throughout the world: why has it come to this? *Can J Anaesth, 53*(6), 540–543.

Wijdicks, E. F. (2012). The transatlantic divide over brain death determination and the debate. *Brain, 135*(Pt 4), 1321–1331.

Wijdicks, E. F., & Pfeifer, E. A. (2008). Neuropathology of brain death in the modern transplant era. *Neurology, 70*(15), 1234–1237.

Wijdicks, E. F., Varelas, P. N., Gronseth, G. S., Greer, D. M., & American Academy of Neurology. (2010). Evidence-based guideline update: determining brain death in adults: report of the Quality Standards Subcommittee of the American Academy of Neurology. *Neurology, 74*(23), 1911–1918.

Yee, A. H., Mandrekar, J., Rabinstein, A. A., & Wijdicks, E. F. (2010). Predictors of apnea test failure during brain death determination. *Neurocrit Care, 12*(3), 352–355.